Climate Change and Sustainable Cities

Climate change has demonstrated, perhaps more than any other environmental concerns, the complexities of the human-nature interrelationship and the need for embedding a far greater environmental consciousness into our social values and norms. A drastic reduction in global greenhouse gas emissions requires a transition to low carbon cities. This demands a better understanding of the interactions between social, technical, and spatial processes which constitute cities.

The aim of this book is to explore these interactions and urge urban planners and other built environment professionals to revisit some of their traditional concepts, methods, and ways of thinking about what constitutes a 'good' city and according to whose priorities. The book brings together nine contributions ranging from broad overviews to sector-specific analysis, paying particular attention to the role of urban planning. Contributors cover climate change mitigation and adaptation, deal with different scales of analysis ranging from international and European to national and city perspectives, and discuss a range of policy sectors including housing, transport, energy, sea level rise as well as pathways for climate policy implementation. The diversity of the contributions is itself a reflection of the multitude of climate change concerns that preoccupy researchers, policy makers and practitioners.

This book was published as a special issue of *European Planning Studies.*

Hugo Priemus is based at the Delft University of Technology, OTB Research Institute for the Built Environment, Delft, The Netherlands.

Simin Davoudi is based at Newcastle University, School of Architecture, Planning and Landscape/Newcastle Institute for Research on Sustainability, Newcastle upon Tyne, United Kingdom.

Climate Change and Sustainable Cities

Edited by
Hugo Priemus and Simin Davoudi

Routledge
Taylor & Francis Group
LONDON AND NEW YORK

First published 2014
by Routledge
2 Park Square, Milton Park, Abingdon, Oxon, OX14 4RN

Simultaneously published in the USA and Canada
by Routledge
711 Third Avenue, New York, NY 10017

Routledge is an imprint of the Taylor & Francis Group, an informa business

British Library Cataloguing in Publication Data
A catalogue record for this book is available from the British Library

ISBN13: 978-0-415-82696-9

Typeset in Times New Roman
by Taylor & Francis Books

Publisher's Note
The publisher accepts responsibility for any inconsistencies that may have arisen during the conversion of this book from journal articles to book chapters, namely the possible inclusion of journal terminology.

Disclaimer
Every effort has been made to contact copyright holders for their permission to reprint material in this book. The publishers would be grateful to hear from any copyright holder who is not here acknowledged and will undertake to rectify any errors or omissions in future editions of this book.

Contents

Citation Information

The chapters in this book were originally published in *European Planning Studies*, volume 20, issue 1 (January 2012), with extra papers from issue 3 (March 2012). When citing this material, please use the original page numbering for each article, as follows:

Chapter 1
Introduction to the Special Issue
Hugo Priemus & Simin Davoudi
European Planning Studies, volume 20, issue 1 (January 2012) pp. 1-6

Chapter 2
Governing Carbon and Climate in the Cities: An Overview of Policy and Planning Challenges and Options
Patricia Romero-Lankao
European Planning Studies, volume 20, issue 1 (January 2012) pp. 7-26

Chapter 3
National Climate Change Adaptation Strategies of European States from a Spatial Planning and Development Perspective
Stefan Greiving & Mark Fleischhauer
European Planning Studies, volume 20, issue 1 (January 2012) pp. 27-48

Chapter 4
Climate Risk and Security: New Meanings of "the Environment" in the English Planning System
Simin Davoudi
European Planning Studies, volume 20, issue 1 (January 2012) pp. 49-70

Chapter 5
Rotterdam: A City and a Mainport on the Edge of a Delta
Han Meyer, Anne Loes Nillesen & Wil Zonneveld
European Planning Studies, volume 20, issue 1 (January 2012) pp. 71-94

Chapter 6
Improving Sustainability in Urban Areas: Discussing the Potential for Transforming Conventional Car-based Travel into Electric Mobility
Bert van Wee, Kees Maat & Cees de Bont
European Planning Studies, volume 20, issue 1 (January 2012) pp. 95-110

Chapter 7

Towards Energy-neutral New Housing Developments. Municipal Climate Governance in The Netherlands
Milly Tambach & Henk Visscher
European Planning Studies, volume 20, issue 1 (January 2012) pp. 111-130

Chapter 8

Improving Energy Efficiency of Social Housing Areas: A Case Study of a Retrofit Achieving an "A" Energy Performance Rating in the UK
Minna Sunikka-Blank, Jun Chen, Judith Britnell & Dimitra Dantsiou
European Planning Studies, volume 20, issue 1 (January 2012) pp. 131-145

Chapter 9

Mediating Low-Carbon Urban Transitions? Forms of Organization, Knowledge and Action
Mike Hodson & Simon Marvin
European Planning Studies, volume 20, issue 3 (March 2012) pp. 421-439

Chapter 10

Local Demonstrations for Global Transitions—Dynamics across Governance Levels Fostering Socio-Technical Regime Change Towards Sustainability
Philipp Späth & Harald Rohracher
European Planning Studies, volume 20, issue 3 (March 2012) pp. 461-479

Introduction

HUGO PRIEMUS* & SIMIN DAVOUDI**

*Delft University of Technology, OTB Research Institute for the Built Environment, Delft, The Netherlands,
**Department of System, Environmental Policy and Planning, University of Newcastle, United Kingdom

Climate change has demonstrated, perhaps more than any other environmental concern, the complexities of the human–nature interrelationship and the fallacy of the modernist assumption about the ability of humans to exploit nature and degrade the environment with little or no consequences. There is now growing awareness that the accelerated rise in global average temperature since the industrial revolution is due to human activities and particularly the burning of fossil fuels. Despite this recognition, the international and national actions taken so far are still inadequate in mitigating the effects of climate change. Efforts are now centrally focused on what is known as the "2°C" challenge. The scientific community is warning that if the global average temperature rises more than 2°C by 2050, the impact of subsequent climate changes would be catastrophic. To keep the rise at this level requires a drastic reduction in global greenhouse gas emissions; at least 50% of the 1990 level by 2050. The "2°C" mitigation challenge calls for fundamental changes to the ways in which energy is produced and used. It also calls for a shift in the ways in which we perceive nature and treat the environment. The transition to low carbon cities is not just a technical agenda; it is a social agenda, too. Technological advances in the use of renewable energy sources, carbon reduction and storage, and increasing energy efficiency of buildings need to be accompanied by behavioural change. For the latter to happen, environmental consciousness needs to be embedded in social values and norms. The same applies to the need for adapting to climate changes that are underway. These have begun to transform the physical geography of the world and confront millions of people (often those who have been less responsible for climate change) with starvation, shortage of water and homelessness. Building adaptive capacity requires both technical knowhow and behavioural change. The scale, urgency, complexity and uncertainty of climate change require concerted actions: by governance at multiple levels, by professionals with different disciplinary backgrounds, and by communities with different experiences. Urban planners and other built environment professionals, among others, are urged to re-visit some of their traditional concepts, methods, and ways of thinking about what constitutes a "good" city (Davoudi *et al.*, 2009).

Climate Change as a New Concern for Spatial Planning

Spatial planning is not, and never has been, an autonomous activity, independent of demographic, economic, political, and technological changes. Technology exerts its influence in a number of ways such as the construction of real estate, transport infrastructure, and hydrological engineering. Climate change is affecting all these areas and not just in a distant future. Its impacts are already manifested through, for example, the rising sea levels that are affecting many parts of the world. These pose new challenges and create new conditions for spatial planning in both urban and rural areas. The focus of this special issue is on the relationships between climate change, energy transition, flood risk, and spatial planning in urban areas. These relationships have only recently become the focus of research and have so far led to more questions than answers. What is clear is that spatial planning in urban areas will never be the same: climate change, energy transition, and sea level rise will have far-reaching impacts on the ways in which cities are planned and managed. There is, therefore, a need to broaden and deepen our knowledge of climate change and its impacts on land use, cities, infrastructure, rural areas, real estate, and the built environment.

There remain many uncertainties, of which some relate to climate change itself: its speed, possible directions and the different ways in which it will manifest itself in various parts of the world. Others concern the impact of climate change on human activities such as the effects of extreme events on cities and infrastructures, the effects of changes in temperature on migration and tourism flows, or the effects of the changing temperature, precipitation and land salination on agriculture. There are also uncertainties about the appropriate technologies and costs of reducing greenhouse gas emissions.

Links Between Cities and Energy Transition

Energy consumption is a major issue in urban areas where fossil fuel emissions are undermining the quality of the environment. CO_2 emissions, which contribute to the greenhouse effect, are generated by industrial and agricultural processes, by the transport sector and by heating in homes and business premises. Therefore, public authorities in urban areas can help lower CO_2 emissions by, for example: promoting renewable energy sources in the energy mix in the electricity grid; reducing car parking facilities in and around the city, increasing access to adequate public transport, and encouraging people to cycle and walk; providing infrastructure for electric cars; promoting sustainability in the energy used for heating space in homes and businesses by introducing combined heat and power generation, geothermal power, and wind and solar energy; and improving the energy efficiency of existing and new buildings.

The general term "energy transition" refers to the gradual changeover from fossil fuels to sustainable (non-fossil) energy sources. This transition is essential because oil supply is expected to peak by the middle of this century. From that point, fossil fuels (coal, oil, natural gas) will no longer be able to meet the rising demand for energy. If action to reduce the consumption of energy in the existing building stock and in new buildings is complemented by efforts to make energy more sustainable, a dual benefit can be obtained. This includes, for example, the use of solar power obtained from solar panels, solar foil, solar boilers, and solar-aware building (passive solar power); wind energy from large and small wind turbines, water as a cooling option, energy from biomass (waste), heat, and cold storage; and district heating.

A great deal is currently expected of "smart grids" hybrid: energy networks at city or regional level, whereby traditional power stations (powered by natural gas or oil and increasingly sustainable energy sources) are linked to homes and businesses that not only consume energy, but also produce it (using geothermal heat, solar cells, biomass and/or wind turbines). Such smart grids can include energy neutral districts and neighbourhoods which produce, distribute, supply, and charge for sustainable energy on a large scale. A combination of creativity, technological innovation, venture capital, entrepreneurship, and political will is needed to accelerate the implementation of these ideas in relation to new buildings and existing stock.

The real estate sector can play a leading role in lowering CO_2 emissions. However, investment in sustainability and energy-saving measures is still limited. A global survey of the environmental performance of real estate funds has exposed huge discrepancies between the illusion and the truth about the implementation of environmental measures by real estate investors and developers. The performance of the top-scoring investors shows that the highest possible score of 100 on the Real Estate Environmental Performance Index is realistic. Real estate investors in Australia, Sweden, and the UK are edging towards the top scores on the index and are therefore leading the field. However, the environmental performance of American investors, in particular, is weak. The same applies to investors in Asia and Southern Europe. So there is plenty of scope for improvement and for lowering CO_2 emissions (Eichholtz *et al.*, 2010).

Links Between Cities and Water

Water has always played a major role in the development of cities (Priemus & Rietveld, 2009). Every city needs reliable sources of drinking water and adequate sewerage systems for waste water. Many cities are located near rivers and shores of seas or lakes. Given the importance of water transport in history, cities with access to water were the places where transport flows were concentrated. Such accessibility also provided good conditions for agricultural production. Many cities developed ports to accommodate the transport flows. However, the silting up of rivers led to the decline of some prosperous cities such as the Flemish city of Bruges in the medieval period. Because of their privileged position, the remaining port cities gradually developed hinterland connections via rail and road, which tended to reinforce their competitive position. This process has led to a strong focus on human settlement systems in locations near places with good water access (including cities such as Singapore, Hong Kong, Rotterdam, Hamburg, Antwerp, and Barcelona). But some port cities where today the maritime activities have shrunk (London, Amsterdam, and Venice) have continued to retain a strong orientation towards their waterfronts with new opportunities for housing, recreation, and tourism.

In metropolitan areas with access to river or sea, water has a strong impact on urban form. With the port activities moving increasingly towards the fringes of the metropolitan area, space is created for new residential, commercial, and recreational activities in central parts of the cities. There are strong indications that the water orientation of these new locations, sometimes linked with heritage aspects of port buildings and infrastructures, adds to their attractiveness for users, creating what Glaeser *et al.* (2001) have coined "consumer city": port heritage and waterfronts adding to the amenities of a city. Thus, apart from the classic transport-related functions, these cities are becoming attractive for residential and tourism functions. Water is also a basic factor for creating green spaces in cities such as parks. Another relevant aspect of water is that it may be a source of energy via the generation of electricity for urban activities when water flows are strong enough. It is clear that, as cities develop, trade-offs will occur between the various land-use types and the role of water. The case of substitution between port activities and residential activities in city centres is an example, but there are many others. One of the reasons to rethink the present approaches to water issues in cities is climate change.

The water orientation of cities is not without its adverse impacts notably the risk of flooding, with New Orleans as a recent tragic example. Old and new port cities are preparing, or have already taken measures, to adapt to the impacts of climate change in terms of increased risk of sea level rise and river flooding (see Meyer and Zonneveld in this issue in relation to the port of Rotterdam). Concerns about climate change-related water management is not just about flooding but also about the degradation of water quality due to, for example, increased salination and pollution from marine transports affecting port cities, in particular. Low water levels in rivers is another source of negative impacts on cities. Cities that depend on electricity and power plants located near rivers may be negatively affected when cooling capacity is reduced due to climate-related drought. Also, inland water transport will be negatively affected when water levels in navigable rivers are low, because low water results in ships transporting smaller loads and leading to higher costs per ton. Low water levels may also have negative impacts on fresh water supply.

Contributions to this Special Issue

The diversity of the papers in this special issue is itself an indication of the multitude of climate change concerns that preoccupy researchers and practitioners. The special issue brings together seven contributions ranging from broad overviews to sector-specific analysis. They deal with climate change mitigation and adaptation and cover different scales of analysis ranging from international (Romero-Lankao) and European (Greiving & Fleischhauer) to national (Davoudi) and city (Meyer & Zonneveld) perspectives. The sector-specific papers focus on housing (Sunikka-Blank *et al.* and Tambach & Visscher) and transport (van Wee *et al.*), drawing on Dutch and UK case studies.

Patricia Romero-Lankao's paper provides an overview of the main climate actors and the multilevel governance of existing mitigation and adaptation responses within an international context. She examines the factors that underpin the gap between the rhetoric and the reality of local climate policy. Drawing on the review undertaken for the 2011 *Global Report on Human Settlements: Cities and Climate Change* (UN-HABITAT, 2011) Romero-Lankao focuses on policy and planning challenges that are involved in the governing of carbon and climate in cities in both global north and global south. A more specific focus on planning for adaptation is presented in the paper by Stefan Greiving and Mark Fleischhauer. Based on an evaluation of existing assessments of climate adaptation strategies and in-depth case studies of nine European countries (undertaken by experts from those countries), they provide a typology of different national approaches to developing spatial strategies for climate change adaptation. Their findings highlight the significance of legal frameworks and political-administrative systems in determining how national adaptation responses are designed and implemented. While the role of spatial planning in delivering climate change adaptation is widely acknowledged, their findings clearly show that this role is largely overlooked in national strategies. They argue that factors that often bring spatial planning to the fore in national adaptation strategies relate mainly to political will and capacity building activities rather than the degree of climate vulnerability or the effectiveness of a particular planning instrument. The focus on spatial planning continues in the paper by Simin Davoudi. However, she adopts a distinctly different approach by focusing on the ways in which climate change agenda has unsettled assumptions about human–nature relationship and introduced new meanings of the environment in the English planning system. Davoudi argues that, while the emphasis on climate change mitigation has reinforced some aspects of the sustainability discourse, the adaptation agenda has introduced a new meaning of the environment as risk. This portrays the environment not so much in terms of assets to be *sustained for* human benefits, but in terms of threats *against which* human wellbeing should be safeguarded. She points out that the current framing of the adaptation agenda in terms of risk, resilience, and security is leading to a resurgence of some outmoded approaches to planning. Han Meyer and Wil Zonneveld explore climate change challenges at the city level with a particular focus on Rotterdam and its relationship with the port at the edge of Rijnmond Delta. They provide a critique of the continuous expansion of the Rotterdam harbour in the light of anticipated sea level rise.

The final three papers in the special issue are sector-specific, focusing on housing and transport and providing case study examples from the Netherlands and the UK. The paper by Bert van Wee, Kees Maat and Cees de Bont challenges the popular assumption that electric cars provide a long-term solution for climate mitigation by addressing the problems associated with the use of fossil fuels. They argue that although electric cars could bring substantial benefits for the environment and energy consumption, their long-term environmental benefits depend on a number of factors such as: the potential rebound effects, the life-cycle effects with respect to the energy used to charge the batteries and the differences between electric cars and their competitors particularly the conventional cars with improved combustion engines. Minna Sunikka-Blank, Jun Chen, Judith Britnell, and Dimitra Dantsiou focus on existing housing which account for the lion share of the housing stock. Making the existing stock energy efficient is central to the ambitious CO_2 reduction targets set by both European Union and the UK government. Based on an on-going case study in the UK, the paper identifies a number of barriers in improving the energy performance rating of social housing and hence questions the feasibility of the current UK policy instruments such as Smart Meters and Feed-in-Tariffs in reducing energy use in social housing. Their findings confirm the points made earlier that the technical and engineering solutions to building construction plays only a limited role in the level of energy used by individuals and households in their homes. Sunikka-Blank *et al.* show that people living in physically identical social housing exhibit a large variety of actual energy used for heating, due to their different lifestyles, occupational patterns, and thermal comfort expectations. The paper confirms the limitations of policy instruments based on the rational choice model. According to this model, people are utility maximizers and their decisions consist of a series of rationally ordered preferences. What tends to be ignored in this model is that people's rationality is bounded both by their limited cognitive capacity to process information and by the social processes in which they operate. The theme of energy efficiency in housing continues in the paper by Milly Tambach and Henk Visscher who focus on energy efficiency in new buildings rather than existing stocks and drawing on the policy measures introduced by the Dutch government. They identify a number of policy instruments which, they argue, are necessary for enabling the Dutch municipal governments to steer new housing developments towards energy neutrality.

References

Davoudi, S., Crawford, J. & Mehmood, A. (Eds) (2009) *Planning for Climate Change: Strategies for Mitigation and Adaptation for Spatial Planners* (London: Earthscan).

Eichholtz, P. M. A., Kok, N. & Bauer, R. (2010) De duurzaamheid van de vastgoedsector [The sustainability of the real estate sector], *ESB*, 95(4584), pp. 266–268.

Eichholtz, P. M. A., Kok, N. & Quigley, J. M. (2010) Doing well by doing good: Green office buildings, *American Economic Review*, 100(December), pp. 2492–2509.

Glaeser, E., Kolko, J. & Saiz, A. (2001) Consumer city, *Journal of Economic Geography*, 1(1), pp. 27–50.

Priemus, H. & Rietveld, P. (2009) Climate change, flood risk and spatial planning, *Built Environment*, 35(4), pp. 425–431.

UN-HABITAT (2011) *Global Report on Human Settlements: Cities and Climate Change*, Geneva: United Nations.

Governing Carbon and Climate in the Cities: An Overview of Policy and Planning Challenges and Options

PATRICIA ROMERO-LANKAO

Resilient and Sustainable Cities Project National Center for Atmospheric Research, Boulder, CO, USA

ABSTRACT *Urban centres play a crucial role in managing global carbon emissions (mitigation) and reducing vulnerability to climate change (adaptation). This paper describes some of the mitigation and adaptation entry points and challenges for city-relevant planning and policy-making posed by the processes defining urban greenhouse gas emissions, vulnerabilities and adaptive capacities. It finds that although many cities are already responding to the climate challenge, existing initiatives are fragmented and a piecemeal rather than a strategic approach is very common. Frequently mitigation and adaptation responses do not address many of the key drivers and determinants involved (e.g. consumption patterns and equity issues determining differentiated access to the determinants of adaptive capacity), nor do they fit with the issues they are intended to address. This is so because climate responses and the issues they are intended to address are multi-scale in nature because most of the processes involved operate at multiple sectoral, temporal and spatial levels. In the face of the complexity of the interconnected processes involved in the relationships between cities and climate change, it is not surprising that local authorities tend to move towards rhetoric rather than meaningful responses.*

1. Introduction

Although climate change has increasingly become recognized as one of humanity's most challenging global environmental issues, it is a deeply local issue as well. In this context, urban centres play a crucial role in managing global carbon emissions (mitigation) and reducing vulnerability to climate change (adaptation). Urban centres are key sources of greenhouse gases (GHGs). They concentrate populations, economic activities and built environments and, thus, are presented with higher risks from floods, heat waves, sea level rise and other hazards that climate change is expected to aggravate. But in spite of the enormous energy challenges, climate risks and vulnerabilities urban areas are facing, these centres will also have opportunities to play pivotal roles in mitigation and adaptation efforts. As hubs of development urban centres have, will, and must become sources of innovations and policy responses to reduce GHG emissions and adapt to the impacts of climate change. The combination of increased challenges with increased opportunities brought to urban areas by climate change can incubate important synergies and resources for creating innovative adaptation and mitigation strategies.

A growing number of scholars have explored how carbon and climate issues are governed at the urban level (Anguelovsky & Carmin, 2011; Betsill & Bulkeley, 2007; Davoudi *et al.*, 2009; Fra Paleo, 2009). Scholarship has shown that these responses are increasingly taking place in a context of multilevel governance ranging from local to sub-national, national and international actors by a redistribution of state-functions *upwards* to international and transnational organizations and institutions; *downwards* to states, regions, urban areas and cities; and *outwards* to NGOs, civil organizations and other non-state actors (Puppim de Oliveira, 2009). Research has found that many carbon and climate-relevant processes of decision-making and institution-building tend to be fragmented, poorly coordinated and ineffective in dealing with the escalating speed and complexity of the mitigation and adaptation challenges at the urban level (Biermann *et al.*, 2009). Furthermore, the responses are seen to be shaped by a series of enabling and constraining external and internal factors such as leadership, competition and political time tables (Anguelovsky & Carmin, 2011).

A more integrated understanding of whether existing mitigation and adaptation responses at the city level are actually addressing the right targets is currently lacking, namely the key emitters and underlying drivers of GHG emissions (Section 2.1) and the determinants of urban vulnerabilities to climate change (Section 2.2). This paper draws on a review commissioned by the UN-HABITAT culminating in the release of *Cities and Climate Change: Global Report on Human Settlements* in April 2011. Key themes in the UN-HABITAT Report are further explored in a special issue of *Current Opinion in Environmental Sustainability* which focuses on some of the multiple linkages between cities and climate change.[1] This paper will use a threefold approach to describe some of the entry points and challenges for planning and policy-making posed by the processes defining urban GHG emissions, vulnerabilities and adaptive capacities (Section 2); to provide an overview of the main climate actors, and the multilevel governance context of existing mitigation and adaptation responses; and to examine some of the key factors explaining the huge gap between the rhetoric and the reality of local climate policy (Section 3). It then closes with some remarks and observations that draw together many of the key themes of the paper (Section 4).

2. What is at Stake

Recent years have witnessed the emergence of new approaches which first focused on alternative energy production, smart urban growth and other policies and strategies aimed at reducing current and future GHG emissions. However, as a new awareness, that mitigation alone cannot prevent all climate-change impacts that have emerged, emphasis has begun to shift to adjusting infrastructures, economic activities and the built environment in order to adapt to future climate challenges (Anguelovsky & Carmin, 2011; Fra Paleo, 2009; Hamin & Gurran, 2009). Regardless of whether these approaches focus on mitigation or adaptation, however, more attention needs to be paid to the key dimensions and determinants of urban emissions and vulnerabilities to climate change. What is at stake, of course, is the very existence of many human institutions, and the safety and well-being of masses of humans themselves, as they are inextricably bound and dependent upon both human systems and ecosystems for their livelihoods at the local, national and global levels. It is only by increasing our understanding of these determinants and dimensions, where they first play out, at the local level, that we can, through effective policy and planning, help to avoid a fast approaching global environmental and economic catastrophe.

2.1 *Proximate and Underlying Drivers of GHG Emissions*

An understanding of the sources and drivers of urban GHG emissions is of utmost importance for planning and action for three reasons. First, each of the main emitters is, directly or indirectly, related to urban functions (Table 1). Each emitter constitutes a universe unto itself, not only in terms of the types of GHGs it generates or the factors explaining differences in the levels and carbon intensities of its emissions, but also in terms of the mitigation opportunities it offers to urban planners and policy-makers.

Energy is by far the most important sector for managing GHG emissions (Kennedy *et al.*, 2009), as the combustion of fossil fuels for electricity generation, heating, cooling, cooking, transportation and industrial production is the major source of GHGs (Table 1). The energy systems that urban areas rely heavily on, and for which, with some exceptions (e.g. Los Angeles), authorities other than local are frequently responsible, are shaped by the quantity of energy used, the energy structure (i.e. renewable forms of energy are preferable) and the quality of the energy (e.g. natural gas is less carbon intensive than coal).

Urban form and density determine differences in the amount of energy used and GHG emitted by *transportation*, a key emission source. The weight of urban density is determined by the dynamics of such processes as the extent of automobile use, the quality of public transit and by land-use planning.

As illustrated in Table 2, Denver's per capita GHG emissions from ground transportation are more than four times greater than those of New York, a much denser city. Similarly, the high level of automobile dependence in Bangkok means that its per capita emissions from ground transportation are twice those of London, a much more affluent city but one with a much more comprehensive public transportation system.

Commercial and residential buildings are key sources of GHG related to on-site combustion of fuels, public electricity use for street-lighting, district heat consumption and the materials used for their construction. Decision-makers need to pay attention to such factors determining emissions by buildings as the needs for heating and cooling (determined by climate conditions, cultural preferences and access to monetary resources), building construction and orientation, the size of the space and the type of fuel used and, lastly, the behaviour of occupants (Dodman *et al.*, 2011a).

Two other emitting sectors are *industry* and *waste*. Because many industrial activities are energy-intensive in their operation, their changing dominance in the economy of cities can make up a big part of their emissions (Dhakal, 2004). Mitigation policies need to address the following factors accounting for the differences in industrial emissions: location, size of the industry, age of the facilities and carbon intensity of their energy sources. Although waste is a small contributor to global emissions, rates of

Table 1. Main emitters and mitigation policies

Emitters			Mitigation policies	
Emitting sector		End use activity	Mitigation areas	Mitigation actions
Energy	Electricity and heat	Residential Commercial	Energy	Carbon content Energy efficiency Renewable energy Educational efforts Demand reduction measurements
	Transportation	Ground air	Transportation	Fleet replacement New low-carbon transport infrastructure Low-carbon transport infrastructure Renewal Fuel switching Energy efficiency Demand reduction measures Demand enhancement measures
	Industry	Industrial	Built environment	Energy-efficient materials and design Alternative energy and water supply systems Energy and water-efficient appliances Demand reduction measurements
	Fugitive emissions	Gas extraction, refining, etc.	Infrastructures	Alternative water and energy supply systems Landfill gas captured Demand reduction measurements Energy and water efficiency/conservation
Industrial processes		Cement, chemicals	Carbon sequestration	Carbon capture from energy generation Tree-planting programmes Restoration of carbon sinks Conservation of carbon sinks Carbon offset schemes
Land-use changes Agriculture		Deforestation, afforestation, etc Livestock, manure	Urban form/development/planning	Land-use planning Compact and green city projects Brown field reuse Small-scale urban renewal

Source: Own based on World Resources Institute: http://cait.wri.org/figures/World-FlowChart.pdf and Bulkeley and Castán Broto (2011).

waste generation have increased in recent years, particularly in middle-income countries that have been experiencing increasing affluence. Waste generation is linked to population, affluence and urbanization, yet as exemplified by Barcelona, London and New York, emissions from waste generation can be greatly reduced by measures such as efficient collection, and technologies for methane capture and recuperation (Dodman, 2009).

The second reason why it is important to understand the sources and drivers of GHG emissions is given by the fact that different, and not always compatible, methodologies for measuring the GHG emissions attributable to cities can be developed depending on definition of spatial boundary and life-cycle perspective (Table 3). These inventories are fundamental planning tools, not only for comparisons and for inter-urban competition and cooperation, but also for identifying and measuring the effectiveness of mitigation measures within cities. Most of the existing GHG inventories only include a single-process perspective and the direct emission sources within the political boundary (Scope 1 method, Table 3). They exclude the indirect and embedded emissions, that policy-makers would need to account for from a life-cycle perspective, i.e. those generated by out of boundary electricity, manufacturing of goods, materials, etc., and food production aimed at satisfying the consumption needs of urban populations (Scope 3 method, see Table 3). For instance, 33% of Chinese emissions (equivalent to 6% of total global CO_2 emissions) in 2005 were due to the production of exports to the rest of the world (Weber *et al.*, 2008). The planning and policy implication of this is that a large measure of the responsibility for the emissions in the generation location should be borne by the urban demographic and economic sectors consuming the goods produced outside their localities. This idea acknowledges the fact that many agricultural and manufacturing activities that meet the needs of urban residents take place outside city boundaries and often in other countries. It sheds a different light on the uneven contribution to the climate-change problem which is at the core of environmental justice issues and the challenges that policy-makers face in finding effective and equitable mitigation solutions. In summary, levels and modes of consumption that are also determined by the marketing

Table 2. Density, ground transportation fuel consumption and GH emissions

	Density (people km^{-2})	Gasoline consumption (ML)	Diesel consumption (ML)	GHG emissions (tonnes CO_2eq per capita)
Bangkok	3606	2741	2094	2.20
Barcelona	16,056	209	266	0.75
Cape Town	1425	1249	724	1.39
Denver	1460	1234	197	6.07
Geneva	1532	260	51	1.78
London	4663	1797	1238	1.18
Los Angeles	905	14,751	3212	4.74
New York City	10354	4179	657	1.47
Prague	2382	357	281	1.39
Toronto	772	6691	2011	3.91

Source: Kennedy *et al.* (2009).

and production strategies of corporations are crucial to understanding the contribution urban areas are making to climate change (Leichenko & Solecki, 2005).

Besides consumption and production patterns, a series of underlying and interlinked factors account for the different contributions of urban areas to GHGs, both within and across countries. The first is *geography,* which alters the climatic situation, altitude and location in relation to sources of energy resources (e.g. hydroelectricity and coal). For instance, latitude determines a city's need for more or less energy to run air conditioning and heating systems within its buildings, industries and dwellings. The *economic base* of a city is the second. "Heavy industrial" cities, such as South Africa, have much higher carbon emissions per capita than financial centres such as the UK (e.g. 19.94 and 6.2 tonnes of CO_2 equivalent yearly for Johannesburg and London, respectively (Dhakal, 2004; Kennedy *et al.*, 2009).

Urban form and density are other determinants (Dodman, 2009; Romero Lankao *et al.*, 2009). Spatially compact and mixed-use urban developments offer other mitigation benefits. First, reduced costs for heating and cooling result from smaller homes and shared walls in multi-unit dwellings (Brown *et al.*, 2008, p. 12). Second, the use of energy systems covers a broader area such as a district, for cooling, heating and power generation as well as lesser line losses related to electricity transmission and distribution (Brown *et al.*, 2008, p. 11). Third, reduced average daily vehicle-kilometres travelled in freight deliveries and by private motor vehicles per capita (VKT). It has been found, for instance, that with all other variables kept constant except density, "the household in a neighbourhood with 1000 fewer units per square mile drives almost 1200 miles more and consumes 65 more gallons of fuel per year more than a household in a higher density neighbourhood" (Brown *et al.*, 2008, p. 12). Urban density is not the only explaining factor, however. Studies on urban transportation have found that besides higher levels of mixed land-use, transit accessibility and ped-

Table 3. Comparison of methods for attributing GHG emissions to cities

Definition	Spatial boundary	Life-cycle perspective	Components
Scope 1	In boundary emissions	Single-process emissions	In boundary: (a) fossil fuel combustion; (b) waste (landfill) emissions; (c) industrial processes and product use; (d) agriculture, forestery and other land use
Scope 2	In boundary electricity use		Out of boundary electricity emissions at powerplant
Scope 3	Out of boundary energy use (and further out of boundary emissions, not included in scope 20	Product life-cycle emissions	All other indirect or embodied emissions: (a) from food and materials consumed in cities (b) upstream of electric powerplants (c) upstream from fossil fuel use

Source: Based on Kennedy *et al.* (2009).

estrian friendliness, attitudes and preferences also influence driving behaviour (Handy *et al.*, 2005). Furthermore, urban density poses a dilemma: while "tailpipe emissions and fossil-fuel consumption are greatly increased with urban sprawl", levels of human exposure to emissions of other pollutants (e.g. NO_x) might actually increase with density if no measures are undertaken to reduce atmospheric emissions (Marshall *et al.*, 2005, p. 284).

It can be misleading to concentrate on urban emissions per capita, as there are very large *differentials within urban centres* related to socioeconomic factors such as income, levels of education and household size. Studies of Indian household CO_2 emissions from energy use and transport (covering rural and urban areas) found that average CO_2 emissions ranged from 335 kg per capita per year for the lowest income class (below 3000 rupees a month) to an average of 1494 kg per capita per year for the highest income class (above 30,000 Indian rupees a month) (Satterthwaite, 2009). Similarly, studies in Mexico City found that energy demand is associated with income and education; energy consumption in poor households was 21% smaller than in high-income households; and energy demand of households with no formal education was 15% lower than in households with education. No correlation was found between gender and energy consumption (Sanchez, 2010).

2.2 *Urban Vulnerability*

Climate variability and change have a variety of implications for urban dwellers, economic sectors and on the network-infrastructures that define the materiality of urban centres, such as energy, waste water and transport systems (Gasper & Ruth, 2011, see Table 3, second column). However, a focus solely on the impacts, hazards and affected sectors does not allow an understanding of the whole array of dimensions and determinants involved in planning for and adapting to climate change, and which have been captured by the notion of urban vulnerability defined as the degree to which a system (e.g. city, population, infrastructure and economic sector) is susceptible to and is unable to cope with adverse effects of a single, or of multiple hazards or stresses such as climate change and political instability (Füssel, 2007). Urban vulnerability is a relative property determined by three interrelated components: (a) the system's exposure (e.g. the degree of severe weather or atmospheric hazards[2] acting upon a particular demographic sector); (b) the system's (or population's) sensitivity or the degree to which a population will respond to a change in the hazard either positively or negatively and (c) the system's ability to adapt or adjust to actual or expected climatic stresses (Romero Lankao & Qin, 2011).

Adaptive capacity, which is different from and influences actual adaptation, is also dynamic and is determined by factors operating at different scales and ranging from age, gender and health conditions, to such assets and entitlements as individual/household resources (e.g. income, health services, air conditioning and/or heating appliances); the extent and quality of infrastructure and public services; and the quality and inclusiveness of governance structures, planning instruments and community organizations that provide or manage safety nets and other short- and longer-term responses (see columns four and five of Table 4).

For urban populations, class and social differentiations are key determinants of income, access to goods and services, health and quality of life (Atkinson, 2007). Rather than being equal, climate risks and vulnerabilities are unevenly distributed in at least two ways. First, a process by which economic elites of urban areas are able to monopolize the best land, and reap the rewards of environmental amenities such as clean air, safe drinking water, open space and tree shade (Harlan *et al.*, 2006; Morello-Frosch, 2002, see Section 3). The second mechanism is given by problem displacement (Castree, 2003), defined as the process by which environmental problems are addressed by disguising and moving them to another space (e.g. untreated water sent to Mezquital an area outside of Mexico City, Romero Lankao, 2010), another time (the future) or even another medium (e.g. GHG emissions). As a result of problem displacement, climate risks are disproportionally borne by the poor, and by marginalized and ethnic minority groups.

Certain processes of urban change have relevance for planning and managing risks from climate and weather hazards at the city level. First, some urban centres are expanding onto low-lying coastal areas (McGranahan *et al.*, 2007) or onto land with agricultural or natural protection potential, or onto land at risk from floods, storms and industrial hazards. Second, large sections of the population, particularly in middle- and low-income countries live in housing constructed informally—with no attention to health and safety standards needed to respond to hazards (UN-Habitat, 2009). Third, although high-density populations are related to lower energy use and reduced GHG emissions (see Section 2), if not adequately provided with such determinants of adaptive capacity as urban infrastructures and services, they can also be related to hazardous conditions such as high concentrations of solid and liquid wastes. This is particularly problematical where sewers, drains and waste collection services are inadequate or absent. Fourth, many provisions for disaster avoidance (e.g. thicker walls), disaster response (access for emergency vehicles) or reducing disaster impacts (readily available open spaces

Table 4. Dimensions and determinants of urban vulnerability

Hazards	Systems	Impacts (changes in)	Determinants of adaptive capacity/ resilience: City wide	Determinants of adaptive capacity/ resilience: Individual level
Sea level rise	Health	Disease	Land-use planning	Age
Temperature	Energy	Mortality	Urban design	Gender
Precipitation	Built environment	Water availability	Transportation	Ethnicity
Heat waves	Economic sector	Air and water quality	Water, sanitation, energy, waste	Migration status
Surges	Demographic group	Economic disruptions	Housing	Income
	Infrastructure	Migration	Social networks	Education
	Transport	Infrastructures damages	Community base organizations	Health condition
	Hinterland	Livelihoods	Policy (emergency) responses	Knowledge, experience
	Ecosystems services		Governance	Savings
				Insurance

Source: Own elaboration.

not at risk from falling buildings) are not possible in crowded low-income settlements. Finally, some characteristics of the urban built environment (e.g. heat-island effect and high levels of atmospheric pollutants) can amplify climate-related risks. For example, the large, impermeable surfaces and concentrations of buildings characteristic of cities can disrupt natural drainage channels and accelerate run-off. The resulting damage from floods can be much more catastrophic if settlements lack drainage or waste collection systems.

Besides being exposed to multiple stresses (De Sherbinin *et al.*, 2007), urban settlements can increase the risk of 'concatenated hazards'. This means that a primary hazard (heavy storm) leads to secondary hazard (e.g. floods creating contamination of water-supplies, or landslides destroying houses and infrastructures). Industrialization, inadequate planning and poor design are key determinants of secondary or technological risks. As illustrated by Bogotá (Colombia), Buenos Aires (Argentina) and Santiago (Chile), the populations of many cities are already at risk from exposure to high levels of pollution, exceeding WHO standards in particulate matter and nitrogen dioxide concentration in the air (Romero Lankao & Gnatz, 2011). Scholars are confirming the possibility that impacts of climate hazards such as heat waves will overlap with pollution events and the urban heat-island effect and compound one another, adding to the complexity of urban disaster risk management (Harlan & Ruddell, 2011).

3. Urban Mitigation and Adaptation Responses

Scholarship on urban responses to climate change has found that local authorities are playing a predominant role in climate-change responses. However, agents such as international organizations, the other tiers of government, grassroots groups, private enterprises, non-governmental organizations and individuals are increasing their participation (Bulkeley & Castán-Broto, 2010). With some overlap, slightly different authorities or sectors are involved in the design and implementation of mitigation from those involved in adaptation actions. Authorities or sectors involved in mitigation often focus on, energy, transportation, built environment, infrastructures and urban planning (Table 1). While those involved in adaptation target systems such as energy, health, infrastructures, urban planning and the management and protection of hinterlands and ecosystem services (Table 4, column 2).

Policies and actions aimed at mitigating or adapting to climate change are increasingly taking place in the multilevel-governance context, and that is often resulting in a fragmented climate-governance. An *ad hoc* governance structure has been created by a patchwork of international institutions (Kyoto Protocol and Hyogo Framework), organizations and networks that are quite distinct in their functions and approaches (e.g. rule setting, information sharing), their constituencies (private and public), their spatial scope (local, bilateral to global), their focus (e.g. mitigation, adaptation, disaster management and development), and their capacity to steer climate-relevant action (Biermann *et al.*, 2009).

These processes are leading to redistributions of functions and responsibilities (e.g. decentralization, privatization of infrastructures), *where urban actors are playing a vital yet often precarious role*. Not only are they involved in supranational initiatives seeking to implement solutions to global warming—e.g. C40, Climate Alliance, they are being affected by structural adjustment programs resulting in decentralization, privatization and devolution of responsibilities and services previously in the hands of national governments (Budds & MacGranahan, 2003). While these changes have sometimes opened opportunities for the creation of more adaptive-governance mechanisms (such as in some Brazilian water basins (Engle & Lemos, 2009)), they have often created additional responsibilities and burdens of service provision that did not come with sufficient resources. Rather than weakening the power of the nation state and creating more local control of these issues, they have often undermined the institutional capacity of local authorities and the resilience of urban centres. The next two sections briefly describe existing mitigation and adaptation responses in the context of the multiple levels of climate change governance.

3.1 *Mitigation*

Urban authorities have developed four 'modes of governing' for mitigation: (a) self-governing given by the capacity of local authorities to govern their own activities, infrastructures, operations and lands; (b) provision or shaping of practice through the delivery of certain—e.g. low-carbon—infrastructure and services; (c) regulation or use of mandates and planning laws as well as oversight and implementation of directives created at other levels of government and (d) enabling, or strategies facilitating, co-ordinating and encouraging action through partnerships with other tiers of government and private actors (Bulkeley & Castán-Broto, 2011) As mentioned earlier, key sectors where mitigation actions are taking place in urban areas include urban development and design, built environment, urban infrastructures, and transport and (Hamin & Gurran, 2009; see Table 1). Action within each of these areas will be briefly highlighted below.

Urban sprawl has become a challenge for governments seeking to direct urban growth in developed and developing countries alike. Various planning strategies have been applied to limit urban expansion, reduce travel and increase the energy efficiency of the urban form: land-use zoning, master-planning, urban densification, mixed use development and urban design standards (UN-Habitat, 2009) have been the theoretical tools used in these efforts. Some of these efforts have taken the form of small- and large-scale regeneration projects (urban renewal), which have taken place primarily in North America, Europe and the former Soviet states (see Table 3). Yet, because the use of land is inherently contentious and politically volatile, local authorities are frequently reluctant to incorporate land-use controls (Berke & Smith, 2009). Within developing countries there are only a few mitigation initiatives that make explicit use of urban form (e.g. compact city planning in Curitiba, Brazil), and when attempted, they are often constrained by a lack of institutional capacity for local governments to effectively implement them. They also have been criticized for their exclusive nature and potential for exacerbating social inequalities (UN-Habitat, 2009).

Economic incentives, regulatory requirements and informational programs have been applied to the *built environment* (Table 1). More recently, a growth in voluntary public–private partnerships has led to an explosion in the range of approaches used including micro generation and new building materials. Nonetheless, built environment projects are primarily taking place within the cities of developed countries (Bulkeley & Castán-Broto, 2011) although they have sometimes taken the form of efforts to help the urban poor (e.g. housing for poor families in Mexico City, Sao Paolo, Bogota and Bishnek, Kyrgyzstan, http://www.holcimfoundation.org and UN-Habitat database). The combination of social and environmental goals may be particularly useful in developing countries for addressing issues such as poverty while creating environmental gains.

Many of the *infrastructure* initiatives have focused on energy efficiency (Table 1) and are primarily driven by such factors as a desire for energy security and financial savings (e.g. Denver, Beijing) and—to a lesser extent—by a desire to take advantage of opportunities to use international instruments such as the Clean Development Mechanism. For instance, methane avoidance and recovery projects make 93% of the CDM projects in urban areas (Disch, 2010). Both drivers may have helped keep these projects economically and politically feasible, but, at the same time may limit their effectiveness in long-term GHG savings as financial savings have sometimes led to increased usage. Because of this, measures must be taken against the rebound effect where gains in efficiency lead to higher energy consumption. Thus, energy efficiency projects need to be coupled with the development of low-carbon renewable fuel sources and efforts to reduce energy consumption.

Climate-change mitigation schemes in the *transport* sector include not only technologically driven actions such as low-carbon transport infrastructure construction and renewal, fleet replacement, fuel switching and enhancing energy efficiency (common in cities of developed and developing countries). They also include demand reduction measures (e.g. disincentives for the use of private motorized vehicles) and demand enhancement measures (incentives for public and other low-carbon modes of transportation such as human powered, Table 1). Furthermore, efforts have been undertaken in cities such as Curitiba, Bogota and Santiago to integrate transportation and land-use through the transit-orientated development idea (UN-Habitat, 2009). The success of the Curitiba model has been attributed to a planning process that strongly integrated land-use and transportation planning, and to continual and long-term attention to the successful implementation of the plan. While the failure of Transantiago resulted not only from lack of integrated transportation planning and poor links to urban planning, but also from a poor understanding of the ways in which people use space and transport in the city (UN-Habitat, 2009).

3.2 *Adaptation*

Some urban authorities around the world[3] are starting to acknowledge the local implications of climate change and are adapting, i.e. introducing technical, physical and regulatory actions in such relevant systems of interest as housing, infrastructures and services, health, emergency response, infrastructures, health and the built environment (Table 4). Some adaptation programs have involved the development of "integrated" climate-change strategies (e.g. Manchester, New York), "stand-alone" strategies (e.g. London, Rotterdam, Durban and Cape Town), and myriad "individual projects" for reducing climate risk (e.g. Mexico City).Yet most of the discussion is still on what cities should do rather than on what they are actually doing to adapt to climate change (Dodman *et al.*, 2011b). Adapting urban areas to climate change is complicated by the fact that it is undertaken at different temporal, spatial and sectoral scales, thus requiring a careful assessment of the different layers involved in land-use planning, housing, emergency responses and their effects on the determinants of urban vulnerability and adaptive capacity at different levels (e.g. city and individual, Table 4). This planning needs to go beyond designing climate action plans or allocating responsibility to departments, such as environmental agencies that are frequently made responsible for managing climate issues (e.g. Durban, Mexico City, Mumbai, Quito, footnote 7), but do not have the decision-making power nor the resources available to address all the dimensions involved. It requires actions not only at the governmental level but also in the ways that bring businesses, grassroots organizations and individuals into the process to perceive and respond to climate change.

Urban populations have long had to cope with a wide range of risks to their economic activities, lives and livelihoods (Wilbanks *et al.*, 2007). Measures such as safe saving schemes, reinforcing homes to withstand extreme weather, diversifying livelihoods, or moving children or valuables to lofts built in the home and reached by climbing ladders (e.g. Dar es Salam & El Salvador see Douglas *et al.*, 2008), become the most frequent type of response to climate hazards. Yet these responses are generally small-scale and do not address the underlying causes of vulnerability and lack of adaptive capacity (see Table 4) Therefore, they could be defined as coping strategies.[4]

Community-based adaptation and grassroots organizations have importance in urban adaptation, for at least two reasons: they help address the limitations or inadequacies of governmental intervention (e.g. in enhancing the creation of saving schemes or the provision of infrastructure and services); they can become an important tool in the enhancement of resilience to extreme weather events whose timing and magnitude are likely to become less predictable. For example, the federations of savings groups, developed by people living in "slums", and active in 15 countries are crucial not only in community-managed risk management, but also in many initiatives that enhance the resilience of the urban poor (e.g. improvement of housing, water, sanitation, drainage and other infrastructures (D'Cruz & Satterthwaite, 2005).

However, community-based adaptation and grassroots organizations are faced with constraints given by the immense cost, energy and time required to construct, develop and maintain the key determinants of resilience for the inhabitants of many cities from middle- and low-income countries, namely infrastructures and services, warning systems, emergency response systems and education (Table 4). Furthermore, grassroots actors should not be idealized, and it should be kept in mind that their extensive involvement in many efforts is difficult (Pelling, 2003). Sometimes, local associations are closely related to the state, or hold private or sectarian interests that distort local action. Change is perhaps most difficult in urban settlements within countries that have experienced strong centralized control or dominance by a local elite. As documented in Guyana (Pelling, 2003), the attempt by the international community to modify urban governance through the funding of community-sponsored development projects runs the danger that local elites or state agents will capture the benefits of grassroots funding.

Cities in high-income countries do not often have large deficits in such determinants of adaptive capacity as infrastructure and services. Most or all of their inhabitants, apart from the homeless, live in buildings that meet building standards; have access to education and are served by piped water supplies, sewers, drains and solid waste collection (Carter, 2011; Zimmerman & Faris, 2011). Therefore, their adaptation responses are frequently easier to design, implement and fund. This does not mean that adaptation effortlessly gets the political support it deserves. Many cities need major upgrades and to climate proof their infrastructures to take account of the likely changes in hazards resulting from climate change. Furthermore, even urban centres in developed countries are not spared from impacts related to lack of or inadequate provision of the some of the determinants of adaptive capacity (Table 4). As illustrated by the 2003 heat wave in European cities or hurricane Katrina in New Orleans in 2005, many climate risks result from differential determinants of adaptive capacity such as those within elderly or poor populations with lack of access to social and family networks, inadequacies in infrastructural protection (e.g. reinforcement of levees) and emergency response systems (D'Ippoliti *et al.*, 2010; Pistrika & Jonkman, 2009).

3.3 *Factors Shaping Mitigation and Adaptation*

The fact that mitigation and adaptation options are becoming part of the political debate within urban centres does, in itself, mean that effective actions can be implemented. A difference exists between the climate rhetoric and proposed options, on the one hand, and successful mitigation and adaptation responses, on the other (see also Davoudi *et al.*, 2009). This can be explained by diverse factors.

The first is the way in which policy issues and response options are framed in terms of local concerns and priorities: e.g. as fitting into existing environmental priorities for improved air quality (Mexico City, Romero Lankao, 2007); or into issues with economic priority such as energy efficiency (for Cape Town & Beijin, see Holgate, 2007; Zhao, 2011, respectively). The need to build more roads to increase the mobility of private vehicles is maladaptive as it leads directly to increases in GHGs (as is the case in many urban areas). However, the mobility issue could be framed to justify urban planning, public transportation systems, etc. Framing is also defined by the priority given to mitigation and adaptation in relation to other and more pressing policy concerns (e.g. economic growth or poverty). For example, the governments of Ilo Peru and Manizales Colombia implemented actions to avoid rapidly growing low-income populations settling on dangerous sites. Although not originally driven by climate-change concerns, these initiatives illustrate how pro-development and pro-poor policies can enhance adaptive capacity (Díaz-Palacios & Miranda, 2005; Velasquez, 2005). Manizales has gone even further by developing an ambitious urban development process, which integrates urban environmental management with local risk management, and more recently, relates climate-change issues within their local development agenda.

Leadership at the individual and organizational levels is also critical in shaping climate-change responses. Some cases (Durban) illustrate the presence of committed individuals (Roberts, 2008); others (Swedish localities, Granberg & Elander, 2007) have an institutional framework within which acting on climate change is supported. External and internal incentives have been another driver for policy action at the urban level. The 2008 Olympic Games were an important catalyst for climate-relevant action in Beijing. They put Beijing in the spotlight and induced it and its national government to make the necessary level of investment to promote sustainable development with special emphasis on energy and air quality (US$12.2 billion) which might be hard to repeat in other Chinese cities (Zhao, 2011).

Urban areas seek to offer business friendly regulatory environment in an effort to attract jobs and tax revenue. However, this practice can create an environmental race for the bottom as regulations protecting the health and well-being of urban inhabitants are cut, thus negatively affecting adaptive action. Similarly, the interests and well-being of the wealthy and powerful segments of an urban population can be given priority over providing even for the basic human needs of less-privileged segments of the population. In this way, access to basic, often life protecting, services and infrastructures can be a question of affluence, as the wealthy segment enjoys a degree of political power which is largely absent in poorer segments.

By putting pressure on decision-makers, extreme weather events and disasters can also foster climate-relevant actions and policies and be the catalyst for changes in the status quo (Carmin & Roberts, 2009; Pelling & Dill, 2010; Vale & Campanella, 2005). Yet, as shown by the slow and limited acceptance of international aid following Cyclone Nargis in Myanmar, fear of political change can lead, at least in the short term, to reinforcing of the status quo and a continuation of the conditions that create high vulnerability to climate hazards.

Institutional capacity is another determinant, given by the autonomy, resources and decision-making power of local authorities in relation to critical sources of GHG emissions and policy sectors, such as transport, land-use planning, infrastructure, building standards, waste and so on (Holgate, 2007; Romero Lankao, 2007). Local authorities that are directly responsible for the provision of waste, transport or energy services (e.g. in many northern European countries) can have much more capacity to reduce emissions (Carter, 2011). They can also get leverage to undertake mitigation actions from supra national powers (e.g. Cities for Climate Protection). Financial resources, whether they are obtained from external or national sources, are a special component of institutional capacity and, as such, both a driver and a barrier to fostering urban responses to climate change (Anguelovsky & Carmin, 2011).

Autonomy, resources and decision-making power of municipal authorities to implement and enforce policies and measures are equally important. In many policy areas, municipal authorities, particularly, but not exclusively, those in developing countries are unable or unwilling to enforce regulations and standards on such key areas as energy, industrial processes and land-use planning (e.g. Latin American cities, Hardoy & Romero Lankao, 2011). Municipal authorities lacking the finances to oversee compliance or to provide even basic services for their constituents are unlikely to invest in both mitigation and adaptation, given the many competing issues on urban agendas. This is especially the case in East Africa where urban infrastructure is already inadequate to cope with the demands of urban populations within the scope of current climate impacts and hazards—and these inadequacies will become even more apparent as a result of climate change. In this context, responses to climate change need to be developed that recognize the synergies between climate change and sustainable urban development (Kithiia, 2011).

The complexity of climate-change science is another factor. Knowledge, be it provided by local universities (as in Cape Town, London and New York) or by both international and local organizations and scientists (such as in Mexico City and Durban), is significant in creating or constraining response capacity as complexities, uncertainties and subjectivities define the development of local inventories and forecasts of GHG emissions; of city-wide assessments of climate hazards, of potential impacts, of underlying vulnerabilities and of potential response options (see for example Carter, 2011).

Another factor is *the inertia* characterizing many of the issues that adaptation and mitigation actions are supposed to address, such as cultural preferences for housing in coastal and forest areas that might be affected by sea level rise or wildfires (e.g. Melbourne Australia). Power plants, refineries and other energy investments of relevance to mitigation and adaptation goals have long lifetimes. Therefore, and beyond the fact that research, development and actions to reduce emissions are urgent, history shows us that it will take decades to centuries to move our current energy system away from its dependency on fossil fuels, the main source of carbon emissions. This is similar to the case of urban infrastructures and urban form. Urban form evolves over time. Yet its rates of change are slow and cannot easily be shaped by design (Bertaud *et al.*, 2009).

4. Concluding Remarks

This paper has explored some of the challenges and opportunities for adaptation and mitigation responses, and briefly characterized the key dimensions and determinants of urban GHG emissions and vulnerability to climate change. These responses and the issues they are intended to address are multi-scale in nature because most of the processes involved operate at multiple sectoral, temporal and spatial levels. Frequently mitigation and adaptation responses do not address many of the key drivers and determinants involved (e.g. consumption patterns and equity issues determining differentiated access to the determinants of adaptive capacity), nor do they fit with the issues they are intended to address. For instance, many of the climatic cause-and-effect relationships are long-term and potentially irreversible and, therefore, require preplanning that includes many sectors and levels and goes beyond the tenure, administrative power and even the lifetime of most current decision-makers and stakeholders. This makes policy decisions in this area particularly difficult, as uncertainties exist in the understanding not only of the drivers but also of the outcomes and impacts of climate change.

Although many cities are already responding to climate challenge, they focus mainly on some components of mitigation (e.g. energy efficiency and transportation), and are only starting to consider adaptation. Many climate programmes do not necessarily address climate concerns, but rather energy security (e.g. Chinese cities) and other development priorities related to economic growth or poverty (e.g. Denver, Manizales). Existing initiatives are fragmented and a piecemeal rather than a strategic approach is very common. Furthermore, notwithstanding the existence of protocols and initiatives to measure mitigation outcomes (e.g. scopes and Project 2°),[5] there is limited information about the individual and collective impact of existing responses, especially when they extend beyond municipal buildings and infrastructure systems or involve behavioural change. As suggested in Section 2, the existence of compatible GHG emissions inventories and of life-cycle approaches that account for issues of environmental justice is fundamental to assessing the impact of these actions.

Most of the urban-mitigation responses (e.g. energy efficiency measures in governmental buildings of Mexico City) target municipal GHG emissions, which make up a small percentage of the overall total, through self-governing mechanisms. This means that too much attention on the government itself may detract resources from the key emitters and broader mitigation challenges faced by a city (e.g. the need to move it away from its dependence on fossil fuels). Seeking to govern climate change through the *provision* of infrastructure and services holds a high potential in those cities where municipal governments may retain ownership or control of infrastructure networks and where basic needs have been met (e.g. Los Angeles, Schroeder, 2011). Because of their targeted and enforceable nature, taxation, land-use planning and other *regulatory* mechanisms (e.g. congestion charge in London) can be very effective in terms of reducing GHG emissions. Yet these are also the least popular approaches adopted by municipal governments, and therefore the most difficult to sustain politically. *Enabling* initiatives such as energy-efficiency clubs with local businesses in Durban, South Africa (Aylett, 2009), have mitigation advantages; they result in (relatively) low upfront economic and political costs; and can increase the transparency and legitimacy of urban governance. However, enabling initiatives are restricted to those who are willing to participate, and cannot be enforced.

While individuals, community-based organizations and local authorities are responding to climate, as yet, too few cities have developed coherent adaptation strategies. The relatively low emphasis on urban adaptation is partly a result of the existing structure of incentives under the climate convention that only recently created an Adaptation Fund. It can also be related to local hindering factors as policy and governance frameworks; societal and cultural inertia; and the fact that getting more widespread attention by city and municipal governments to climate-change adaptation requires clear and detailed risk assessments (knowledge), as well as a better understanding of how adaptation measures can serve and be integrated not only into disaster risk reduction, but also into components of the development agenda such as land-use planning, water access, sanitation and housing.

In summary, in the face of the complexity of the interconnected processes involved in the relationships between cities and climate change, it is not surprising that local authorities tend to move towards rhetoric rather than meaningful responses. The complex interplay of these factors may, in fact, enhance or constrain a local government's ability to mitigate the consequences of climate change or to respond to its inevitable impacts. What is important is that we understand what is at stake and find ways of fostering successful mitigation and adaptation in these dynamic systems.

Acknowledgments

This work is supported by the National Science foundation (NSF) HPCC, 9139, 7785, 7726. Any opinions, findings and conclusions, recommendations or omissions expressed are those of the author and do not necessarily reflect the views of NSF. I want to thank Daniel Gnatz and the reviewers for their valuable suggestions and input to this paper.

Notes

1. A commissioned Background Paper and three chapters of the 2011 UN-Habitat Report on Cities and Climate Change (Romero Lankao, 2008, 2011; Romero Lankao & Gnatz, 2011; Romero Lankao *et al.*, 2011) as well as a 2011 special issue of *Current Opinion in Environmental Sustainability* on "Cities in transition: Transforming urban centers from hotbeds of GHG emissions and vulnerability to seedbeds of sustainability and resilience".
2. According to Turner *et al., (*2003) hazards are "...threats to a system, comprised of perturbations and stress (and stressors), and the consequences they produce".
3. E.g. London, UK, Boston, New York, Los Angeles, USA, Hamilton and Wellington; Durban and Cape Town, South Africa, Esmeraldas, Ecuador; Kampala, Uganda; Beijing, China; Maputo, Mozambique; Mexico City and Sorsogon City, the Philippines (see Anguelovsky & Carmin, 2011; Carter, 2011; Hardoy & Romero Lankao, 2011; Hunt & Watkiss, 2007; Liu and Deng, 2011; Roberts, 2008; Satterthwaite *et al.*, 2007; UN-Habitat, 2009; Zimmerman & Faris, 2011).
4. Although coping and adaptation refer to actual responses that depend on many determinants of adaptive capacity, they differ in some respects. While adaptation has a long-term perspective, coping mechanisms have a short-term perspective and can increase vulnerability or undermine sustainable adaptation strategies over the long term. Coping includes temporal changes to behaviour or employment of techniques that allow urban actors to deal with immediate hazard events, but then return to their original practices after the pressing danger has passed (Gallopin, 2006).

5. Project 2° is a collaboration between the Clinton Climate Initiative, Microsoft Corporation, Autodesk and ICLEI, which "allows cities to establish a baseline on their greenhouse gas emissions, manage inventories, create action plans, track the effectiveness of their emissions reduction programs, and share experiences with each other". http://www.project2degrees.org/Pages/Default.aspx Viewed on 07/30/2010

References

Anguelovsky, I. & Carmin, J. (2011) Something borrowed, everything new: Innovation and institutionalization in urban climate governance, *Current Opinion in Environmental Sustainability*, 3(3), pp. 169–175.

Atkinson, W. (2007) Beck, individualization and the death of class: A critique, *The British Journal of Sociology*, 58(3), pp. 349–366.

Aylett, A. (2009) *Changing Perceptions of Climate Mitigation Among Competing Priorities: The Case of Durban, South Africa Case Study Prepared for Cities and Climate Change*. Retrieved from Cities and Climate Change. Available at http://www.unhabitat.org/grhs/2011 (accessed 17 December 2009).

Berke, P. & Smith, G. (2009) Hazard mitigation, planning, and disaster resilience: Challenges and Strategic choices for 21st Century, in: U. Fra Paleo (Ed.) *Building Safer Communities. Risk Governance, Spatial Planning and Responses to Natural Hazards*, pp. 1–18 (Amsterdam: IOS Press).

Bertaud, A., Lefevre, B., & Yuen, B. (2009) GHG emissions, urban mobility and efficiency of urban morphology: a hypothesis, in: *Fifth Urban Research Symposium, Cities and Climate Change: Responding to an Urgent Agenda*, Marseille, France.

Betsill, M. & Bulkeley, H. (2007) Looking back and thinking ahead: A decade of cities and climate change research, *Local Environment*, 12(5), pp. 447–456.

Biermann, F., Betsill, M. M., Gupta, J., Kanie, N., Lebel, L. & Liverman, D. (2009) *Earth System Governance—People, Places, and the Planet*, The Earth System Governance Report 1, Bonn, IHDB Report 20.

Brown, M. A., Southworth, F., & Sarzynksi, A. (2008, May). *Shrinking the Carbon Footprint of Metropolitan America* (Washington, DC: Brookings Institute).

Budds, J. & MacGranahan, G. (2003) Are the debates on water privatization missing the point? Experiences from Africa, *Asia and Latin America Environment and Urbanization*, 15(2), pp. 87–113.

Bulkeley, H. & Castán-Broto, V. (2010) *Global Cities and Climate Change Experiments: A Database of 100 Cities* (Durham, UK: Durham, University) (available from the authors on request).

Bulkeley, H. & Castán Broto, V. (2011) Chapter 5: Climate change mitigation responses in urban areas, in: N. D. Mutizwa-Mangiza, B. C. Arimah, I. Jensen, E. A. Yemeru & M. K. Kinyanjui (Eds) *UN-Habitat, 2011 UN-Habitat Report on Cities and Climate Change*, p. 104 (London: Earthscan).

Carmin, J. & Roberts, D. (2009) Government institutions and innovations in governance for achieving climate adaptation in cities, in: *Fifth Urban Research Symposium, Cities and Climate Change: Responding to an urgent Agenda*, Marseille, France.

Carter, J. (2011) Climate change adaptation in European cities, *Current Opinion in Environmental Sustainability*, 3(3), pp. 193–198.

Castree, N. (2003) Commodifying what nature? *Progress in Human Geography*, 27(3), pp. 273–297.

Davoudi, S., Crawford, J. & Mehmood, A. (Eds.) (2009) *Planning for Climate Change: Strategies for Mitigation and Adaptation for Spatial Planners* (London: Earthscan).

D'Cruz, C. & Satterthwaite, D. (2005) *Building Homes, Changing Official Approaches: The Work of Urban Poor Federations and Their Contributions to Meeting the Millennium Development Goals in Urban Areas*, Working Paper 16, p. 80 (London: IIED).

De Sherbinin, A., Schiller, A. & Pulsipher, A. (2007) The vulnerability of global cities to climate hazards, *Environment and Urbanization*, 19(1), pp. 39–64.

Dhakal, S. (2004) *Urban Energy Use and Greenhouse Gas Emissions in Asian Megacities. Kanagawa* (Japan: Urban Environmental Management Project Institute for Global Environmental Strategies).

Díaz Palacios, J. & Miranda, L. (2005) Concertacion (reaching agreement) and planning for sustainable development in Ilo, Peru, in: S. Bass, H. Reid, D. Satterthwaite & P. Steele (Eds) *Reducing Poverty and Sustaining the Environment*, pp. 255–279 (London: EarthScan Publications).

D'Ippoliti, D., Michelozzi, P., Marino, C., De'donato, F., Menne, B., Katsouyanni, K., Kirchmayer, U., Analitis, A., Medina-Ramón, M., Paldy, A., Atkinson, R., Kovats, S., Bisanti, L., Schneider, A., Lefranc, A., Iñiguez, C. & Perucci, C. A. (2010) The impact of heat waves on mortality in 9 European cities: Results from the EuroHEAT project, *Environmental Health*, 9(37), p. 8.

Disch, D. (2010) A comparative analysis of the 'development dividend' of Clean Development Mechanism projects in six host countries, *Climate and Development*, 2(1), pp. 50–64.

Dodman, D. (2009) Blaming cities for climate change? An analysis of urban greenhouse gas emissions inventories, *Environment and Urbanization*, 21(1), pp. 185–201.

Dodman, D., Huq, S. & Satterthwaite, D. (2011a) Chapter 3: The contributions of urban areas to climate change, in: N. D. Mutizwa-Mangiza, B. C. Arimah, I. Jensen, E. A. Yemeru & M. K. Kinyanjui (Eds) *UN-Habitat, 2011 UN-Habitat Report on Cities and Climate Change*, pp. 33–64 (London: Earthscan Publications).

Dodman, D., Huq, S. & Satterthwaite, D. (2011b) Chapter 6. Climate change adaptation responses in urban areas, in: N. D. Mutizwa-Mangiza, B. C. Arimah, I. Jensen, E. A. Yemeru & M. K. Kinyanjui (Eds) *UN-Habitat, 2011 UN-Habitat Report on Cities and Climate Change*, pp. 129–162 (London: Earthscan Publications).

Douglas, I., Alam, K., Maghenda, M. A., Mcdonnell, Y., Mclean, L. & Campbell, J. (2008) Unjust waters: Climate change, flooding and the urban poor in Africa, *Environment and Urbanization*, 20(1), pp. 187–206.

Engle, N. & Lemos, M. C. (2009) Unpacking governance: Building adaptive capacity to climate change of river basins in Brazil, *Global Environmental Change*, 20(1), pp. 4–13.

Fra Paleo, U. (2009) Building Safer Communities. Risk Governance, Spatial Planning and Responses to Natural Hazards (Amsterdam: IOs Press).

Füssel, H. (2007) Vulnerability: A generally applicable conceptual framework for climate change research, *Global Environmental Change*, 17(2), pp. 155–167.

Gallopin, G. (2006) Linkages between vulnerability, resilience, and adaptive capacity, *Global Environmental Change*, 16(3), pp. 293–303.

Gasper, R. & Ruth, M. (2011) Social and economic impacts of climate change on the urban environment, *Current Opinion in Environmental Sustainability*, 3(3), pp. 150–157.

Granberg, M. & Elander, I. (2007) 'Local governance and climate change: Reflections on the Swedish experience', *Local Environment: The International Journal of Justice and Sustainability*, 12(5), pp. 537–548.

Hamin, E. M. & Gurran, N. (2009) Urban form and climate change: Balancing adaptation and mitigation in the U.S. and Australia, *Habitat International*, 33(3), pp. 238–245.

Handy, S., Xinyu, C. & Mokhtarian, P. (2005) Correlation or causality between the built environment and travel behavior? Evidence from Northern California, *Transportation Research Part D*, 10(6), pp. 427–444.

Hardoy, J. & Romero Lankao, P. (2011) Latin American cities and climate change: Challenges and options to mitigation and adaptation responses, *Current Opinion in Environmental Sustainability*, 3(3), pp. 158–163.

Harlan, S. & Ruddell, D. (2011) Climate change and health in cities: Impacts of heat and air pollution and potential co-benefits from mitigation and adaptation, *Current Opinion in Environmental Sustainability*, 3(3), pp. 126–134.

Harlan, S. L., Brazel, A. J., Prashad, L., Stefanov, W. L. & Larsen, L. (2006) Neighborhood microclimates and vulnerability to heat stress, *Social Science and Medicine*, 63(11), pp. 2847–2863.

Holgate, C. (2007) Factors and actors in climate change mitigation: A tale of two South African cities, *Local Environment: The International Journal of Justice and Sustainability*, 12(5), pp. 471–484.

Hunt, A. & Watkiss, P. (2007) *Literature Review on Climate Change Impacts on Urban City Centres: Initial Findings* (Paris, France: ENV/EPOC/GSP).

Kennedy, C., Ramaswami, A., Carney, S. & Dhakal, S. (2009) Greenhouse gas emission baselines for global cities and metropolitan regions, in: *Fifth Urban Research Symposium, Cities and Climate Change: Responding to an urgent Agenda*, Marseille, France.

Kithiia, J. (2011) Climate change risk responses in East African cities: Need, barriers and opportunities, *Current Opinion in Environmental Sustainability*, 3(3), pp. 176–180.

Leichenko, R. M. & Solecki, W. D. (2005) Exporting the American dream: The globalization of suburban consumption landscapes, *Regional Studies*, 39(2), pp. 241–253.

Liu, J. & Deng, X. (2011) Impacts and mitigation of climate change on Chinese cities, *Current Opinion in Environmental Sustainability*, 3(3), pp. 188–192.

Marshall, J., McKone, T., Deakin, E. & Nazaroff, W. (2005) Inhalation of motor vehicle emissions: Effects of urban population and land area, *Atmospheric Environment*, 39(2), pp. 283–295.

McGranahan, G., Balk, D. & Anderson, B. (2007) The rising tide: Assessing the risks of climate change and human settlements in low elevation coastal zones, *Environment & Urbanization*, 19(1), pp. 17–37.

Morello-Frosch, R. P. (2002) Environmental justice and regional inequality in Southern California: Implications for future research, *Environmental Health Perspectives*, 110(2), pp. 149–154.

Pelling, M. (2003) *The Vulnerability of Cities: Social Resilience and Natural Disaster*, p. 212 (London: Earthscan).

Pelling, M. & Dill, K. (2010) Tipping points for change in the adaptation of sociopolitical regimes, *Progress in Human Geography, Progress in Human Geography*, 34(1), pp. 21–37.

Pistrika, A. K. & Jonkman, S. N. (2009) Damage to residential buildings due to flooding of New Orleans after hurricane Katrina, *Natural Hazards*, 542(2), pp. 413–434.

Puppim de Oliveira, J. A. (2009) The implementation of climate change related policies at the subnational level: An analysis of three countries, *Habitat International*, 33(3), pp. 253–259.

Roberts, D. (2008) Thinking globally acting locally—institutionalizing climate change at the local government level in Durban, South Africa, *Environment & Urbanization*, 20(2), pp. 521–537.

Romero-Lankao, P. (2007) How do local governments in Mexico City manage global warming? *Local Environment*, 12(5), pp. 519–535.

Romero-Lankao, P. (2008) *Urban Areas and Climate Change: Review of Current Issues and Trends*, Issues paper prepared for 2011 UN-Habitat Report on Human Settlements, Boulder: NCAR.

Romero-Lankao, P. (2010) Water in Mexico City: What will climate change bring to its history of water-related hazards and vulnerabilities? *Environment and Urbanization*, 22(1), pp. 157–178.

Romero-Lankao, P. & Gnatz, D. (2011) Chapter 2. Cities and the International Climate Change Framework, in: N. D. Mutizwa-Mangiza, B. C. Arimah, I. Jensen, E. A. Yemeru & M. K. Kinyanjui (Eds) *UN-Habitat, UN-Habitat Report on Cities and Climate Change*, pp. 17–32 (London: Earthscan Publications).

Romero-Lankao, P. & Gnatz, D. (2011) Introduction. Urbanization and the challenge of climate change, in: N. D. Mutizwa-Mangiza, B. C. Arimah, I. Jensen, E. A. Yemeru & M. K. Kinyanjui (Eds) *UN-Habitat, UN-Habitat Report on Cities and Climate Change*, p. 1–16 (London: Earthscan Publications).

Romero-Lankao, P. & Qin, H. (2011) Conceptualizing urban vulnerability to global climate and environmental change, *Current Opinion in Environmental Sustainability*, 3(3), pp. 142–149.

Romero-Lankao, P., Tribbia, J. L. & Nychka, D. (2009) Testing theories to explore the drivers of cities' atmospheric emissions, *Ambio*, 38(4), pp. 236–244.

Romero-Lankao, P., Wilbanks, T. & Gnatz, D. (2011) Conclusion, linkages between climate change responses and policy directions, in: N. D. Mutizwa-Mangiza, B. C. Arimah, I. Jensen, E. A. Yemeru & M. K. Kinyanjui (Eds) *UN-Habitat, UN-Habitat Report on Cities and Climate Change*, pp. 163–183 (London: Earthscan Publications).

Sanchez, L. (2010) Inequality and household energy consumption in Urban Mexico. Paper presented at 2010 PAA Meeting, April, Dallas, TX (unpublished manuscript).

Satterthwaite, D. (2009) The implications of population growth and urbanization for climate change, *Environment and Urbanization*, 21(2), pp. 545–567.

Satterthwaite, D., Huq, S., Pelling, M., Reid, A. & Romero Lankao, P. (2007) *Building Climate Change Resilience in Urban Areas and among Urban Populations in Low- and Middle-income Nations*, Commissioned by the Rockefeller Foundation, International Institute for Environment and Development (IIED) Research Report, 112 pp.

Schroeder, H. (2011) *Climate Change Mitigation in Los Angeles, US*, Unpublished case study prepared for the Global Report on Human Settlements 2011, Available at http://www.unhabitat.org/grhs/2011 (accessed 5 September 2009).

Turner, B. L., II, Kasperson, R. E., Matson, P. A., McCarthy, J. J., Corell, R. W., Christensen, L., Eckley, N., Kasperson, J. X., Luers, A., Martello, M. L., Polsky, C. Pulsipher, A. & Schiller, A. (2003) A framework for vulnerability analysis in sustainability science, *Proceedings of the National Academy of Science*, 100(14), pp. 8074–8079.

UN-Habitat (2009) *Global Report on Human Settlements 2009, Planning Sustainable Cities* (London: Earthscan).

Vale, L. & Campanella, T. (Eds.) (2005) *The Resilient City: How Modern Cities Recover from Disaster* (New York: Oxford University Press).

Velasquez, L. (2005) The Bioplan: Decreasing poverty in Manizales, Colombia, through shared environmental management, in: S. Bass, H. Reid, D. Satterthwaite & P. Steele (Eds) *Reducing Poverty and Sustaining the Environment*, pp. 44–72 (London: Earthscan Publications).

Weber, C., Peters, G., Guan, D. & Hubacek, K. (2008) 'The contribution of Chinese exports to climate change' Energy Policy, *Energy Policy*, 36(9), pp. 3572–3577.

Wilbanks, T., Romero Lankao, P., Bao, M., Berkhout, F., Cairncross, S. & Ceron, J.-P. (2007) Chapter 7: Industry, settlement and society, in: M. L. Parry, O. F. Canziani, J. P. Palutikof, P. J. van der Linden & C. E. Hanson (Eds) *IPCC WGII Fourth Assessment Report*, pp. 357–390 (Cambridge, UK and New York, NY, USA: Cambridge University Press).

Zhao, J. (2011) *Climate Change Mitigation in Beijing, China*, Unpublished case study prepared for the Global Report on Human Settlements 2011, Available at http://www.unhabitat.org/grhs/2011 (accessed 2 August 2009).

Zimmerman, R. & Faris, C. (2011) Climate change mitigation and adaptation in North American cities, *Current Opinion in Environmental Sustainability*, 3(3), pp. 181–187.

National Climate Change Adaptation Strategies of European States from a Spatial Planning and Development Perspective

STEFAN GREIVING & MARK FLEISCHHAUER

Institute of Spatial Planning, TU Dortmund, Dortmund, Germany

ABSTRACT *The aim of this article is to give an overview and systematic characterization of different national approaches to developing strategies of climate change adaptation from a spatial planning or regional development perspective, respectively. Based on this analysis, recommendations are made for the implementation of the Territorial Agenda of the European Union. The central research hypotheses addressed in this article are: (1) climate change impacts in Europe are distributed differently in European regions which influence the design of national adaptation strategies (NASs) in Europe as well as the planning-related fields of action; (2) the legal framework and the political-administrative system significantly determine how national adaptation responses are designed and by which institutions they are implemented and (3) spatial planning has the potential to play an important role in climate change adaptation due to its integrative, cross-sectoral character. In order to discuss the hypotheses, a meta-evaluation of already existing assessments of climate adaptation strategies was carried out, as well as nine in-depth country studies (Finland, France, Germany, Greece, Hungary, Poland, Spain, The Netherlands and the UK). The results show that spatial planning is only given minor attention in the assessed analyses and national strategies and that it is rather a matter of political willingness and capacity building than particular instruments or a high climate change vulnerability if spatial planning, however, plays an important role in an NAS.*

1. Introduction

Planning and planners are generally regarded to be responsible and capable to reduce vulnerability and to develop climate mitigation and adaptation capacities against the impacts of climate change (Stern, 2006; IPCC, 2007). Also, the World Bank Report "The Global Monitoring Report 2008" which deals with climate change and the Millennium Development Goals concludes that the development of adaptive urban development strategies is a fundamental field of action for dealing with the challenges of climate change (World Bank, 2008).

Also the European Union (EU) White Paper "Adapting to climate change: Towards a European framework for action" (European Commission, 2009a, p. 4) explicitly relates to spatial planning and territorial development, respectively: "a more strategic and long-term approach to spatial planning will be necessary, both on land and on marine areas, including in transport, regional development, industry, tourism and energy policies". Further, the EU Commission aims to define guidelines and exchange best practice "to ensure that account is taken of climate change impacts when implementing the Environmental Impact Assessment (EIA) and Strategic Environmental Assessment (SEA) Directives and spatial planning policies" (European Commission, 2009a, p. 13).

In the EU Territorial Agenda (BMVBS, 2007a, p. 7), it is stipulated under priority 5 that

> joint trans-regional and integrated approaches and strategies should be further developed in order to face natural hazards, reduce and mitigate greenhouse gas emissions and adapt to climate change. Further work is required to develop and intensify territorial cohesion policy, particularly with respect to the consequences of territorially differentiated adaptation strategies

As the Territorial Agenda is at a stage of implementation, this article aims to recommend spatial planning-related approaches to climate change adaptation.

Since 2005, several national adaptation strategies (NASs) to climate change have been developed and adopted in the EU Member States. The role of sectoral planning, such as water management or landscape planning, is highlighted throughout the NASs, but the role of spatial planning or territorial development often remains unclear—in contrast to the role that is dedicated to spatial planning in the aforementioned policy documents.

1.1. *Objective and Research Hypotheses*

The main objective of this paper is to provide an overview and systematic characterization of different national approaches to climate change adaptation from a spatial planning or regional development perspective, respectively.[1]

> *Hypothesis 1*: Climate change impacts in Europe are distributed differently in European regions due to the variation in climate stimuli. The regional differences influence the design of national adaptation strategies in Europe as well as the planning-related fields of action.

Cross-regional vulnerability analyses in general differentiate between regional types of climate change exposure due to climate stimuli (e.g. Greiving *et al.*, 2010) and the sensitivity of regions towards climate change. Both in combination determine potential climate change impacts. Furthermore, adaptive capacity and coping capacity can be differentiated regionally (Massey & Bergsma, 2008). However, some common spatial patterns of climate change impacts can be identified, i.e. regions with similar climate change impacts. Thus, one research question is whether regional differences in climate change impacts influence the role of spatial planning in NASs.

Hypothesis 2: The legal framework and the political-administrative system significantly determine how national adaptation responses are designed and by which institutions they are implemented.

Although the policies of the community shape national policies more and more, the specifics of the different Member States mainly characterize the design of national policies. In some of the EU Member States, for example, a new development is legally allowed when it conforms to the regional plan. This so-called regulatory function of spatial planning is known under the term "conforming planning" in the international discourse on planning theory (Larsson, 2006; Rivolin, 2008). In most of the EU Member States and thus also in the majority of country studies (Section 3), however, the so-called development function dominates at the regional planning level which is discussed under the term "performing planning". This planning type is characterized by non-binding programmatic and/or strategic statements. Potential projects are then evaluated against the question whether they support the implementation of the programme or strategy. Furthermore, there are—if at all—only partially binding effects for the subordinated local level. The research question in this respect is whether the planning system has an influence on the role of spatial planning in NASs.

Hypothesis 3: Spatial planning has the potential to play an important role in climate change adaptation due to its integrative, cross-sectoral character.

As climate change impacts occur in specific places and might cause conflicts with land use and regional development, there is a need to find territorially relevant answers to this challenge. At the same time, spatial planning tools offer a variety of approaches to reduce the negative impacts of climate change, leading to the question what the role of spatial planning is in fact.

1.2. *Methodological Approach*

The analysis was carried out in two methodological steps.

1. In a first step, 11 cross-country assessments of climate change adaptation strategies were analysed by a meta-evaluation (Section 1 and Table 1). The main research interest was to answer the question which role spatial planning and spatial development play in climate change adaptation strategies. This step had an explorative character, and the results were used to structure the assessment design for the second step of the analysis.
2. In a second step, nine country studies were carried out and analysed (Section 3). The countries were to cover all parts of Europe including the different regional climate change impacts and were also to represent the variety of spatial planning systems. The main research of the second step was to verify or falsify the aforementioned hypotheses. The countries were assessed in a cross-sectional analysis from the perspective of spatial planning and development, respectively.

Finally, recommendations for new spatial planning approaches and tools are made with special respect to the implementation of the Territorial Agenda of the EU, taking into account the empirical findings as well as the latest literature to the issue.

Table 1. Overview of the selected comparative climate change adaptation studies

Authors and title of the study	Context	Central contribution to this study
European Commission (2009b): Towards a European framework for action impact assessment	This impact assessment to the White Paper on Climate Change Adaptation included comparative statements	Integration of climate proofing into the strategic environmental assessment
Ribeiro *et al.* (2009): Design of guidelines for the elaboration of Regional Climate Change Adaptation Strategies	Commissioned by the EU Commission. Compared national and regional adaptation strategies	The high importance of the involvement of stakeholders; NASs to legitimize regional strategies
Swart *et al.* (2009): Europe adapts to climate change. Comparing national adaptation strategies	The study was carried out by the non-governmental PEER network in order to allow insights into the different strategies of European countries	The need to integrate climate change adaptation into existing planning instruments and communicate the benefits of spatial planning as coordinator of sectorial planning
Mickwitz *et al.* (2009): Climate policy integration, coherence and governance	The study was carried out by the non-governmental PEER network in order to reflect upon the political implementation of adaptation strategies	The divergence between the claim and the factual relevance of spatial planning for the adaptation process
Brooks *et al.* (2009): Prioritizing climate change risks and actions on adaptation—A review of selected institutions, tools, and approaches	The study was carried out for the Canadian "Policy Research Initiative" and analyses institutional adaptation concepts with the aim to prioritize risks from climate change impacts and adaptation options	The relevance of prioritizing risks from climate change impacts for the development of adaptation options
CIRCLE ERA-Net (2009): Outcomes of the 1st International CIRCLE Workshop on Climate Change Adaptation	CIRCLE aims at establishing a European research platform for climate impact research, adaptation and vulnerability and organizing the exchange between climate research and politics	Focus on synergies instead of conflicts between climate change mitigation and adaptation
Schuster (2008): Klimaanpassungsstrategien in europäischen Nachbarländern (Climate change adaptation strategies in Germany's neighbouring countries)	Core questions were the identification of adaptation measures with spatial planning relevance from countries neighbouring Germany and the transferability of such measures to Germany	Particular spatial planning-related adaptation measures for specific vulnerabilities

Meister *et al*. (2009): Schwimmende Häuser und Moskitonetze: Weltweite Strategien zur Anpassung an den Klimawandel—Nationale Strategien und Projektbeispiele (Floating houses and mosquito nets: Worldwide strategies for the adaptation to climate change—national strategies and sample projects)	Guiding question of the study was how the analysed countries prepare for the impacts of climate change. The study derives success factors for the development and implementation of adaptation strategies that shall provide a first orientation in this new policy field	Success factors for adaptation strategies (cross-cutting topic, multitude of stakeholders, long-term perspective, actions under uncertainty and local implementation in a global context)
Roggema (2009): Adaptation to Climate Change: A Spatial Challenge	This study is part of the book "Adaptation to Climate Change: A Spatial Challenge". Guiding question of the study is which role spatial planning can play in adaptation strategies	The variety of adaptation strategies and their influencing factors
Fleischhauer *et al*. (2006): Natural Hazards and Spatial Planning in Europe	Guiding question of the study is which role spatial planning (at the regional and local levels) can play in dealing with natural hazards	Limits of formal spatial planning instruments; spatial planning as one of many actors
Massey and Bergsma (2008): Assessing adaptation in 29 European countries	The study was commissioned by the Netherlands Environmental Assessment Agency in order to provide an overview of the state of European adaptation approaches	Regionally differentiated vulnerability and specific adaptation strategies and measures

Source: authors' own construct.

2. A Review of Comparative Studies on Climate Change Adaptation Strategies

The results of the meta-analysis comparative studies on climate change adaptation strategies served as the basis for the research design of the country studies (Section 3). Furthermore, its conclusions are integrated into the recommendations of Section 4. Table 1 provides an overview of the analysed studies.

The studies analysed in general have a broad scope and only in three cases a clear spatial planning or territorial development perspective (Fleischhauer *et al.*, 2006; Schuster, 2008; Roggema, 2009; the latter primarily focuses on natural hazards in general and not on NASs).

The analysis shows that spatial planning is only of marginal interest. Instead, the studies focus on the state of implementation of NASs and the included sectors. Sometimes, success factors, weaknesses and knowledge gaps are identified. Particular projects or the role of specific planning instruments was of less interest. The studies that specifically focus on the role of spatial planning in adaptation strategies, however, are an exception in this respect: Schuster (2008) and Roggema (2009) as well as the study of Fleischhauer *et al.* (2006).

There is a need to distinguish between the role of planning relating to relevant topics (land use and urban design) and spatial planning institutions as actors. For example, land use is often mentioned as a cross-cutting topic. Land-use-related adaptation measures mentioned in the analysed studies (Table 2) such as recycling of brownfield areas, reduction of land consumption or the adaptation of buildings to higher temperatures or extreme weather events often dedicate a certain role to spatial planning. However, spatial planning is generally seen to play only a subordinated role—with the exception of the studies from Schuster (2008) and Roggema (2009). Interestingly, spatial planning is often categorized as one of many sectors, and the authors of the studies hardly recognize the coordinating function of spatial or regional planning. Roggema regards spatial planning as an important tool in any adaptation process due to its integrating character. The studies also show that all assessed countries have unique approaches to climate change adaptation. Generally, a focus on spatial planning is missing. This has a lot to do with the fact that a cross-sectoral coordination function and the mandate to regulate and safeguard spatial development as known, for example, in Germany are not common in most of the European countries.

The majority of the studies highlight the importance of a sound information basis. However, they emphasize that in spite of the existing uncertainty concerning the dynamics and impacts of climate change, there is already sufficient knowledge at hand to act now, especially because of the long-term nature of climate change. Therefore, thinking in scenarios is needed as well as the often postulated focus on no-regret or low-regret measures.

Continuity in planning and implementation and a periodic evaluation of the objectives of strategies and measures are therefore essential. The potential importance of impact assessments—also for prioritizing climate change impacts and monitoring—is often highlighted and discussed, especially in the context of strategic environmental assessment.

All studies assign a key role to the involvement of stakeholders in any kind of adaptation process. Often, the development of cross-departmental, cross-sectoral and cross-level governance structures with stakeholder networks is discussed, resulting in so-called multi-level governance structures. Nevertheless, many studies conclude that only little attention is paid to communication processes at present and especially in the first phases of an adaptation process.

Table 2. Strengths and weaknesses of spatial planning in the context of climate change adaptation

Task	Milestones	Potential of spatial planning	Description
Assessment of long-term consequences	Assessment and appraisal of climate change impacts on the human-environmental system	Fair	This is possible based on regional impact studies that planning has to have at hand. A strength of comprehensive planning is the traditionally integrated view of different change processes (demography, economy, environment and climate)
Climate proofing	Identification of interaction between land uses and the changing climate	Good	Such assessments can easily be integrated in the strategic environmental assessment which is obligatory for any spatial plan or programme
	New guiding principles (such as "resilience")	Good	The concept of resilience is almost in line with existing planning principles such as decentralized concentration and could therefore easily be adopted by planning practice
	Avoiding non-adapted developments	Good	This is within the focus of planning which is very much about future developments. The effectiveness of actions depends partly on the existing regulatory framework (zoning instruments)
	Adaptation of existing spatial structures (settlements and infrastructure)	Poor	Any adaptation of existing structures is hardly possible through regulatory planning due to the given private property rights. What is needed are incentives and good practices aiming at convincing the private landowners
Disaster prevention	Assessment of frequency and magnitude of extreme events (exposure)	Poor	That is clearly a task for specialized authorities such as water management where spatial planning usually does not have any competence
	Keeping disaster-prone areas free of further development	Good	At least conforming planning systems have regulatory zoning instruments at hand. Keeping free of areas prone to extreme events is thereby possible
	Differentiated land use according to the given risk	Fair	Almost possible, but not effective with regard to existing settlement structures
	Adaptation of existing building structures	Fair	Almost impossible through regulatory measures due to property rights. Suitable approaches are based on incentives and communication with all stakeholders
	Relocation/retreat from threatened areas	Poor	Again, this is in conflict with property rights. Full recompensation is normally needed, which is mostly impossible due to the lack of public financial resources. It is possible in areas with shrinking population where the existing building stock will be (partly) deconstructed based on planning strategies (see, e.g. Eastern Germany)

Source: authors' own construct.

All studies agree that climate change adaptation is a cross-cutting issue that touches a broad range of policy fields. It remains open, however, how the horizontal and vertical coordination could be guaranteed. The studies mention that only the integration of climate change adaptation into the different sectoral policies is very important. The potential importance of spatial planning in this respect is not recognized. This may be due to the fact that most of the reviewed studies do not consider spatial planning at all and that the cross-sectoral coordination function of regional planning only exists in a few countries.

The importance of prioritizing adaptation measures, e.g. on the basis of cost–benefit analysis, is often mentioned. The conflicts between climate change mitigation and adaptation, in contrast, only play a minor role in the reviewed studies.

3. Analysis of Country Studies

The following section presents the country studies that were carried out exclusively for the aforementioned study (Greiving & Fleischhauer, 2010), which is the basis for this article. The analysis of the NASs was not only restricted to one specific strategic document, but describes the whole set of adaptation policies at the national level. The NAS thus encompasses strategic frameworks, action plans, implementation policies, etc. Further, the state of the NASs as of early 2010 was analysed no matter if the strategy was formally adopted, conceptualized or—as in the case of Greece and Poland—only existing in preliminary papers. The analysis was carried out along a common commented table of contents for all country studies.

3.1. *Overview of Results*

All in all, there are large differences between the NASs with regard to the objectives, the substantive scope and the spatial focus. The sectors that are addressed are nevertheless similar throughout the assessed countries.

With respect to the role of spatial planning, the adaptation strategies can be grouped as follows.

1. *Greece and Poland*. An NAS does not exist. The role of spatial planning is officially not yet defined. Nonetheless, a preliminary judgement of the current situation is possible based on scientific literature and informal documents (Karaczun *et al.*, 2009; Wassenhoven *et al.*, 2009).
2. *Finland and Spain*. The existing NASs acknowledge spatial planning as a sector and/or relevant actor. A detailed description of its role is missing.
3. *France, Germany and Hungary*. The existing NASs acknowledge spatial planning as a sector and relevant actor. The particular role of planning is described, but not translated into planning practice yet.
4. *The Netherlands and the UK*. The existing NASs as well as action plans acknowledge spatial planning as a key sector and relevant actor. The particular role of planning is described and (partly) implemented in planning law and practice.

In general, the development of the NASs happened under the leadership of the respective ministry for the environment, often in cooperation with national cross-sectoral authorities (France, Germany, the UK, Greece, Poland and Spain). In the Netherlands, the spatial planning and environmental ministry is the coordinating institution. In some of the countries, the strategy was developed in an intensive cooperation process with several ministries (France, the Netherlands and Spain).

The existing NASs are informal, non-binding documents of a very different character. For example, in France, the NAS aims at setting up a frame for the subsequent implementation strategy, but the prescriptions of the adaptation strategy are not binding. In the Netherlands, the NAS serves as an orientation both for the substantive implementation (adaptation agenda) and for the organizational implementation (inter-provincial agreement). In Germany, the adaptation strategy aims more at delineating the responsibilities and competences between the federal government, the federal states, the regions and the local level. In the UK, the strategy is very much focused on the stakeholders that are relevant for the implementation of the strategy and directly addresses public and private organizations.

In all countries, water management is the dominant sector concerning climate change adaptation as water-related issues are directly touched by a range of climate change impacts such as flooding (river flooding and flash floods) or water shortages (dryness and droughts) and trigger other social and economic impacts. In Finland, this is the sector that is most advanced in terms of implementation. In the Netherlands, naturally, the whole area of climate change adaptation is closely connected to the water sector—only to give two examples from the country reports. In all assessed countries, the cross-sectoral character of climate change adaptation is considered. Consequently, multiple sectors and different types of regions are addressed in many strategies.

The costs and benefits of adaptation measures are discussed at different levels. One aspect is the comparison and analysis of costs and benefits, and the other is the general question of how adaptation measures can be funded. Although these questions seem to be of central importance, they are only marginally touched in the NASs, most likely due to the methodologically open questions of how to estimate costs and benefits of adaptation measures. Above all, the question of economic or financial aspects is generally only asked in the context of the subsequent implementation of the strategies.

Some strategies such as the Dutch and the Hungarian offer specific tools for vulnerability assessment or comprehensive performance indicators regarding adaptation at the local level. Others are still vague with respect to the specific contribution of spatial planning to adaptation. Even in those countries where climate change adaptation has been acknowledged as an important issue for spatial planning, it has not been implemented yet into planning practices broadly.

There are only a few integrative concepts which link adaptation and other key challenges of spatial development such as demographic change, settlement development and economic growth. This is a particular weakness of the existing strategies because in planning practice, it is necessary to weigh adaptation measures against other relevant issues and interests. Therefore, those adaptation measures are preferred that embody synergies, but not conflicts with other interests. It is also astonishing that synergies as well conflicts between adaptation and mitigation are rarely discussed by the strategies (i.e. dense settlement patterns that, on the one hand, lower CO_2 emissions by stimulating public transport, but, on the other hand, also increase exposure to the urban heat island effect). What is widely missing is an operationalization of commonly accepted strategies (like the "no-regret" strategy) for the application in practice.

Moreover, there is an open question concerning synergies and conflicts between climate change mitigation and adaptation measures, on the one hand, and adaptation measures regarding climate change and other important trends (demographic change, globalization, etc.), on the other hand. Only few statements can be found in this regard in the country studies. In the German adaptation strategy, synergies and conflicts between mitigation and adaptation are explicitly mentioned, and it is emphasized that adaptation measures shall not be implemented at the expense of mitigation. The Hungarian adaptation strategy points at potential synergies and conflicts between adaptation and mitigation without going into detail (Vajdovich, 2010).

In the Netherlands, all programmes, plans and projects have to be developed in such a way that they are climate-proof, i.e. that they are resilient to changing conditions that they might be confronted with in the future (Fleischhauer, 2010). This climate proofing is part of the Strategic Environmental Assessment (for programmes and plans) and the Environmental Impact Assessment (for projects) in the Netherlands.

A broader concept has been implemented in Hungary: an ex-ante Territorial Impact Assessment, which addresses the potential impacts of development strategies across the full range of spatial functions and land uses, including but not exclusively dealing with climate proofing (Vajdovich, 2010).

Depending on the stage of implementation of the NASs, adaptation to climate change gradually becomes a guiding principle for all spatially relevant decisions. However, in most countries, monitoring of climate adaptation activities vis-à-vis the NASs has not been specified yet. Indicators for monitoring the implementation of the adaptation plans—that shall also be used in the regions—have only been developed in Spain (Saurí, 2010) and the UK (Davoudi, 2010).

While several countries are clearly on this way (Germany, France and Finland), the UK has already reached a stage where the climate change adaptation principle became binding by law (Her Majesties Stationary Office, 2008) and has been implemented as NI188 performance indicator into the Local Government Performance Frameworks. However, the future of local area agreements and the national indicator set may be reviewed in the context of the new government's "localism" agenda. Nonetheless, the Climate Change Act remains to be in force. Moreover, the need for further attention to adaptation was highlighted in the Environmental Audit Committee's 2010 report on Adapting to Climate Change (House of Commons, 2010), which suggested that the Government must make adaptation and mitigation more central to the planning system. New developments should only be permitted if they are suited to future climates and support the overall resilience of the built environment.

3.2. *Intermediate Findings*

The uncertainty about future climatic conditions is seen as the main challenge for designing strategies and particular measures for climate change adaptation. This is particularly relevant for spatial planning. For weighing-up the different interests, detailed, spatially highly resoluted information is requested by planners about the impact of climate change. This, but also the persistent existing settlement and infrastructure patterns call for flexible strategies. This is outlined in several adaptation strategies which focus on a programmatic framework, on the one hand, and stakeholder involvement, discourse-based processes and empowering regional and local communities to cope with climate change, on the other hand.

Adaptation activities are still very much disaster-driven because extreme events influence the perception of climate change while less obvious, creeping changes are often disregarded by the public. This shows that climate change adaptation such as natural hazard mitigation, too, is often pushed aside by other issues on the agenda until a certain event happens. Examples, also pointed out in the country reports, are the 2003 heat wave which had major repercussions in France, floods in Germany, Greece, Poland and the UK and forest fires in Greece. Issues which are related to climate change adaptation are in many cases communicated through disaster management, which is usually better established than spatial planning. Therefore, adaptation is at least indirectly addressed in those countries that do not yet have an adaptation strategy, such as Greece and Poland.

The local level is regarded as the most important level for implementing national and regional adaptation strategies and related amendments to planning laws. This is due to local responsibilities for urban development and building permissions, but also to the fact that the population, in general, has more trust in local authorities.

There are some aspects that are relevant in all analysed countries, such as budget constraints and a lack of coordination among the respective actors. However, in some countries such as Greece and Poland, cultural beliefs and historical experiences (with the socialist period) play a relevant role for the existing aversion against any kind of long-term planning, including adaptation planning.

In regard to climate change adaptation, all assessed countries foresee similar tasks for spatial planning:

- adapting to extreme events either by keeping areas free from further development, which are already prone to natural disasters and might even be more endangered in the future, or by ensuring development that is adapted to the future hazards;
- developing building guidelines that are adapted to the changing climate;
- planning settlement patterns which are characterized by sufficient density, but also include open spaces for limiting the effects of urban heat islands;
- managing urban water resources in a sustainable way which considers the consequences of heavy rainfall but also potential water shortcomings during summer time.

The existing interactions between adaptation and other change processes such as demographic change and economic transition are often disregarded by the NASs. However, on the ground, these challenges are often considered to be more relevant for local development strategies. Moreover, it is stipulated that adaptation has to be communicated as a new issue for planning and not as an issue which dominates all other existing issues. Starting points have to be identified in order to illustrate connections to already established planning principles such as compact settlement structures, which can be seen as a multifunctional principle useful for mitigation and adaptation but also for avoiding urban sprawl and improving the competitiveness of an urban area.

However, planning is able to address these tasks when developing new settlement plans, but Europe is dominated by persistent settlement structures, cultural landscapes and infrastructures which have been developed over centuries and which are most sensitive to climate impacts. Preventive actions, carried out by spatial planning, are under these circumstances less effective than that in countries which are still growing rapidly in terms of population and the built environment. Regulatory planning is not well suited to enforce adaptation in already built-up areas due to private property rights and obligatory compensation payments for using land for public purposes. Thus, all strategies regard financial incentives and communication as relevant measures because the acceptance by private property owners needed for implementing climate change strategies. Under these conditions, climate change governance becomes crucial for the success of any adaptation strategy.

Mickwitz *et al.* (2009, p. 60) came to the following conclusion:

> While the need for co-ordination and integration across sectors, scales and levels is growing, the capacities to respond are frequently shrinking [...]. While it is generally recognised that the role of spatial planning for climate mitigation and adaptation should be strengthened, the practice is not very well developed as yet.

. This statement can be backed by this article. Only in the Netherlands, the national government considers spatial planning a key player.

Concluding, there is no visible connection between the attention paid to spatial planning and the type of planning system or vulnerability to climate change of the respective country. Consequently, other factors such as political priorities of a national government might determine the relevance of spatial planning for adaptation.

In view of the weak performance of current planning practices in climate change adaptation, the key question is how to strengthen the role of spatial planning. This is important for the implementation of the Territorial Agenda of the EU and also national, regional and local adaptation strategies and will be discussed in the following.

4. Recommendations

The following recommendations are drawn on the cross-analysis of the nine country reports. They are addressed to the implementation of the Territorial Agenda of the EU. At the invitation of the Portuguese presidency, an Informal Ministerial Meeting on Territorial Cohesion and Regional Policy was held in 2007. The Ministers met to discuss and reach a common view on the issues related to territorial cohesion. Action 2.1 is about cooperation with the European Commission to use the activities of the Territorial Cohesion and Urban Matters thematic working groups (TCUM) to analyse key dossiers, policies and legislation, at EU and Member State level, in the light of the Territorial Agenda and the Leipzig Charter. TCUM 1 focuses on Climate Change. These recommendations contributed to TCUM 1 and guided the political debate about the territorial dimension of adaptation to climate change in Europe.

The background document "The Territorial State and Perspectives of the European Union" (BMVBS, 2007b, p. 36) outlined several considerations for policies in the section "Risk Management in Relation to Climate Change". Not all of these have been considered yet in the analysed NASs. This is particularly valid for the proposed improvement of coordination between disaster risk and climate adaptation strategies, for sequential strategies such as the proposed triangle "resistance—resilience—retreat", but also for the principle of a polycentric spatial development which—according to the Territorial Agenda—can balance patterns of vulnerability in Europe. Other elements such as integrated vulnerability assessments and stakeholder involvement concepts have been widely applied in different NASs.

4.1. *Cross-sectoral Character of Adaptation as Multi-level Governance*

Climate change adaptation calls for a cross-sectoral approach because of the variety of impacts on different sectors and the interdependences between impacts and response strategies. Mickwitz *et al.* (2009) argue for a prominent role for comprehensive spatial planning in this regard. This has to be prominently addressed and outlined by the Territorial Agenda. Moreover, following the example of the Netherlands, the role of the ministries responsible for spatial planning has to be strengthened for coordinating the implementation of the NASs.

The implementation of adaptation strategies calls for a broad involvement of all societal groups in order to guarantee the legitimacy of actions. This relates to the uncertainty of climate change impacts and the fact that there is no scientific evidence of how, when and to which extent climate change impacts exactly will occur. Political decisions based on uncertain knowledge need a broad mandate from all social groups. In particular, quantitative goals have to be justified because they are of a normative character.

Adapting the existing settlement patterns to the challenges of climate change can be seen as the main challenge for spatial planning operating in the context of existing private property rights. Incentives and more inclusive discourse-based approaches are needed, which can be characterized as "climate governance". This aspect is regarded as an important success factor by recent literature on adaptation strategies (European Commission, 2009a; Meister *et al.*, 2009; Ribeiro *et al.*, 2009; Swart *et al.*, 2009).

Table 2 summarizes to what extent spatial planning can contribute to cope with climate change. Three main areas can be distinguished: assessment of long-term consequences, climate proofing and prevention of disasters triggered by climate change.

4.2. *Integration of Adaptation to Climate Change into Spatial Visions*

Spatial planning can contribute to adaptation by providing visions for resilient spatial structures, as argued, for example, by the German Adaptation Strategy (DAS). However, spatial planning is requested to coordinate different spatial functions, land uses and related interests. Therefore, any spatial vision has to consider a broad range of issues. However, uncertainty characterizes all types of planning processes which face grand challenges such as demographic change and globalization. Hence, there is a general need for resilience—not only in regard to climate change adaptation. Consequently, planners should—together with civil society organizations—agree on spatial visions that are characterized by the following elements.

- *Efficiency*. Efficient spatial structures produce and deliver products and services on less space with fewer resources (energy and natural resources). This mitigates greenhouse gas emissions and minimizes the exposition of sensitive land uses against extreme events. In doing so, protected resources could be used only if needed in case of unpredicted or unpredictable developments.
- *Diversity*. Diverse settlement structures (mixture of infrastructures, buildings and open spaces) contribute to sensitivity reduction because different land uses have different sensitivities in regard to a particular extreme event or creeping changes in temperature and precipitation.

- *Redundancy*. The functionality of an urban system could be better ensured if its main elements are redundant and could replace each other. Therefore, the traditional planning principle of bundled infrastructure (roads, telecommunication, water supply, etc., using the same space or development corridors) becomes questionable.
- *Robustness*. Robustness of infrastructure, buildings and vegetation provides potential damages from the impacts of extreme weather events but also creeping changes.

An updated Territorial Agenda of the EU should communicate resilience in such a way in order to make it more illustrative and understandable for planning practice.

4.3. *Climate Proofing*

Climate proofing is a concept that aims at integrating principles of climate change adaptation into spatial planning. A first definition was offered by Kabat *et al.* (2005, p. 283):

> Climate proofing does not mean reducing climate-based risks to zero—an unrealistic goal for any country. The idea is to use hard infrastructure to reduce risks to a quantified level, accepted by the society or economy. This risk can be further combated by "softer" measures, such as insurance schemes or, as a last resort, evacuation plans. Such climate proofing should be driven by opportunities for technological, institutional and societal innovations, rather than purely by fear of the negative effects of climate change.

The conceptual frame of climate proofing was originally created in the Netherlands (Kabat *et al.*, 2005) and has been implemented into the NAS and related policy documents (Fleischhauer, 2010). However, similar concepts are part of the UK adaptation strategy (see Planning Policy Statement 1, Davoudi, 2010).

Climate proofing aims at resilient spatial structures. This is the main focus of the Dutch adaptation strategy, which has been developed from a spatial planning perspective. Thus, spatial planning in the Netherlands should be "climate-proof". In the UK, all national planning documents and investments as well as regional and local planning have to regard adaptation in order to lower the existing vulnerability to climate change and to build up climate change resilience. In practice, this means that all spatial goals for new developments, but also existing spatial structures, have to be adapted to the needs of adaptation to climate change. Consequently, not only the tools of spatial planning, but also other spatially relevant policies such as infrastructure investments and energy have to be considered, as outlined particularly in the UK. For that purpose, a climate-sensitive development checklist was developed (Davoudi, 2010).

Up to now, a climate vulnerability assessment is required on all spatial levels only in the UK and is to be repeated every 5 years (see Climate Change Act). In Hungary, the scope of the so-called "Territorial Impact Assessment" is much broader and includes not only a vulnerability assessment, but also all other spatially relevant potential impacts of programmes and plans.

Continuous monitoring is needed for the implementation of climate proofing, in particular for controlling the impacts of plans or programmes. In the case of deviations from the objectives, counter measures have to be taken. In the UK, the process-oriented performance indicator on adaptation (NI 188) aims at monitoring the progress of local governments in regard to adaptation to climate change.

4.4. *Climate Proofing as Relevant Criterion for Distribution of Spatially Relevant Funds*

Spatial development is to a large extent influenced by financial incentives, e.g. coming from European structural funds as well as national and regional funds. The disbursement of these funds is based on different criteria for the selection of projects for funding. However, until now, these criteria relate mainly to economic and partly to social and environmental aspects while climate proofing does not play a considerable role. As argued by the European Commission (2009b, p. 131 ff), adaptation should be considered when updating existing funding schemes.

- The Member States should mainstream adaptation when they revise their current Operational Programmes.
- Rural Development Funds could support agriculture's contribution to protection and enhancement of green infrastructure, such as developing green corridors that facilitate species migration in response to climate change, involving wider land-use planning, including trans-boundary collaboration and supporting the development and adoption of more resilient crops.
- A climate impact assessment could become a mandatory condition for projects receiving significant EU funding (for instance, tourism investments should avoid spending money for projects that are unsustainable in the longer term, e.g. ski facilities in low land Alpine areas).
- Energy efficiency and the use of renewable energies for housing projects should become an important factor for funding—like those implemented in the respective German national funding scheme.

Another issue relates to synergies between adaptation and urban consolidation programmes. As already outlined, the adaptation of existing settlements is the major challenge for Europe. There are several programmes which aim at consolidating cities in industrial transition or with a shrinking and ageing population (i.e. in Germany, the programme "Stadtumbau" (urban restructuring) funded by the Federal Ministry for Transport, Housing and Urban Development). These programmes should consider adaptation needs as a selection criterion for funding: decisions about consolidation measures should be based not only on economic tools such as cost–benefit analysis, but also on a climate impact assessment. In doing so, even retreating from built-up land becomes a realistic option for reducing the vulnerability to extreme events (e.g. floods but also heat waves).

4.5. *Integrative Concepts, Addressing Adaptation Together with Other Spatially Relevant Challenges*

Mitigation and adaptation of climate change and demographic change have to be seen as core challenges for an integrated, sustainable urban development (ARGEBAU, 2008). The main challenge for planning is to identify synergies and conflicts between the different problems a city is confronted with. On this basis, an integrated development strategy has to be developed and implemented. This integrative approach is widely missing in the analysed adaptation strategies and related documents. However, the comprehensive, cross-sectoral competences of spatial planning clearly enable planners to tackle the different challenges altogether. By integrating and connecting adaptation with other spatially relevant issues, it could be much better communicated to the public. This communication at the local level is about particular adaptation measures, whose suitability is based on the local circumstances and cannot be generalized.

Consequently, the suitability of any adaptation measure has to be clarified in each single case individually. This also explains why integrative concepts are missing in the NASs, while calling for specific local and regional information on which kind of synergies and conflicts might be related with a particular set of adaptation options. Spatial development more than ever requires the integration of the manifold social, ecological and economic aspects. In addition, the increasing complexity and uncertainty confront decision makers with particular difficulties. Until today, only a few and mostly large cities have developed concepts for facing climate change. Especially, medium- and small-sized municipalities lack (human) resources and capacities for developing appropriate adaptation strategies. Advice could be given by a decision-support-system (DSS) which

- systematically compiles information about synergies and conflicts which could be related with a particular combination of measures;
- thereby supports the development of an integrated, almost conflict-free strategy;
- offers good practices and background documents about adaptation options.

Interesting examples for such a DSS are the UK Adaptation Wizard (http://www.ukcip.org.uk/index.php?Itemid=273&id=147&option=com_content&task=view) and the German Stadtklimalotse (www.stadtklimalotse.net).

4.6. *Parametric Governance*

During the last years, the understanding of government has changed from a traditional approach where the state itself provides all kinds of public services to a model which is characterized mainly by private service providers, while the state only guarantees the quality of the services which are of public interest (Johnson, 1991; Neary, 2001).

One important reason can be seen in the lack of financial resources and the alteration of the classical hierarchical government: the addressees have control over important resources (e.g. information, creativity, knowledge, etc.) not only in connection to risk-related issues. Furthermore, less and less standardized solutions can be purported to manage the given situation (e.g. of existing risk, etc.). Therefore, the state switches to bilateral or multilateral forms of cooperative governance, where the addressees concentrate on new, appropriate solutions and targets as well as output-oriented governance (summarized under the term “agreement on objectives”). This approach aims at governance via parameters or requirements. The modality of achieving the objectives remains in the hands of the addressees (Cools *et al.*, 2003).

To sum it up, the “agreement on objectives” approach is characterized by flexibility and a variety of solutions that lead to a greater motivation and implementation of all involved actors (Cools *et al.*, 2003). In comparison to traditional input-oriented decision-making, the “agreement on objectives” approach offers more flexibility, which is needed for coping with climate change where conclusions cannot necessarily be drawn on the basis of past events.

Uncertainty is considered to be one of the main characteristics of climate change. Under this condition, strategies for spatial planning are needed, which anticipate uncertainty. A suitable way would be based on the so-called "robustness principle": uncertainty would be tackled by avoiding one optimal final state as planning objective, but seeking a set of sub-optimal achievable final states. Planning decisions on the way to these states will only be taken if they are indispensable, while further steps remain open as long as possible. Decisions will be taken as soon as the given uncertainty is tackled by more precise information (in this case about future climatic conditions and the related vulnerability; Friend & Jessop, 1973). An application for planning purposes would be based on so-called "if–then relations" (Greiving, 1998): designations for further settlement development have to be prioritized: only those projects can be immediately realized that are not in question due to existing uncertainty about future (climatic) conditions. Other projects have to be checked for their implementability from time to time considering new knowledge about future (climatic) conditions. They can only be realized if a net social and economic benefit becomes clearly visible. Such a strategy depends on continuous monitoring of the environment, which is, however, already required according to article 10 § 1 SEA Directive.

The following steps which have partly been implemented by the UK National Indicator N 188 for Local Authorities and Local Authority Partnerships (DEFRA, 2010; LRAP, 2010) characterize such a concept of parametric governance:

1. *setting the scene*: undertake a climate vulnerability assessment as information basis every 5 years. This assessment should be integrated into the Strategic Environmental Assessment;
2. agreement on quantitative objectives for vulnerability reduction by lowering the current sensitivity and/or building adaptive capacities;
3. implementation of these objectives through an action plan, which has to be updated based on repeated vulnerability assessments;
4. monitoring of success through qualitative process indicators that are part of annual self-assessments (classes, for example, from 0 to 4 expressing the performance of the actor with regard to the following aspects):
5. definition of responsibilities,
   - identification of relevant actors,
   - assessment of current status of the environment,
   - assessment of vulnerability to climate change,
   - development of an adaptation strategy,
   - setting-up of an action plan,
   - implementation of adaptation measures,
   - monitoring and update of the strategy;
6. connecting the agreement on objectives with spatially relevant funds (as implemented in France, see Angignard, 2010): the allocation of funds should be based on an agreement on objectives between regional and local actors.

4.7. *Spatial Planning Institutions as Relevant Actors for Strengthening Adaptive Capacities*

The relevance of adaptive capacity was recognized by most of the analysed studies and all NASs. However, the role of spatial planning in this context remains widely open.

This paper, along the lines of previous research of the ATEAM project (Schröter *et al.*, 2004), considers adaptive capacity to consist of three parts: awareness, ability and action. Knowledge and awareness as determinants of adaptive capacity play an important role for identifying vulnerabilities in relation to climate change and suitable adaptation measures. In order to move from awareness to action, ability is necessary, which consists of technology and infrastructure within a given society. Finally, the ability to achieve action is supported by economic resources and institutions that enable a society to carry out the adaptation measures that have been defined.

Although this is up to now not really translated into practice, spatial planning is in principle very much relevant for all three parts of adaptive capacity.

- *Awareness*. Awareness building particularly among private landowners calls for trust in those public authorities that are conducting communication campaigns. In most cases, the population trusts most in local authorities that they have direct contacts with. These are typically the municipal planning departments, which regularly involve the public in the setting-up of plans and programmes.
- *Ability*. Due to its cross-sectoral nature, spatial planning is able to coordinate and weigh different adaptation options against each other. This is obvious when looking for technology-based adaptation options (e.g. improvement of coastal defence systems) which often imply conflicts with other land uses. Further, resilient settlement structures are considered important for improving adaptive capacities to deal with extreme events (e.g. resistant infrastructure and building stock) as well as creeping changes (e.g. through improved storage capacity of a sewage system).
- *Action*. Spatial planning is relevant for the allocation of facilities related to coping capacity such as fire brigade stations, hospitals, etc.

These potential contributions of spatial planning to adaptive capacity building should be highlighted by a revised Territorial Agenda.

5. Conclusion

There is a wide consensus that adaptation is a cross-cutting issue. Some authors argue for a prominent role for comprehensive spatial planning in this regard, but this has not been implemented yet.

The findings from the analysis lead to the following discussion of the hypotheses which were formulated at the beginning of the study.

> *Hypothesis 1*: Climate change impacts in Europe are distributed differently in European regions which influence the design of national adaptation strategies in Europe as well as the planning-related fields of action.

Interestingly, there is only partly a clear interconnection visible between the climate change vulnerability profile of a particular country and the key sectors addressed by its adaptation strategy. Almost the same sectors are considered as relevant (i.e. water resources, biodiversity and agriculture) throughout the countries. However, they are considered as important for different reasons which become clear when looking at particular adaptation measures at regional and local levels. However, climate change adaptation is mainly considered to be relevant for spatial planning, amazingly by those countries such as the UK, Finland or the Netherlands which are characterized as being only moderately vulnerable compared with others in Eastern and Southern Europe (Spain, Greece and Poland), where less attention is paid yet to the role of spatial planning and adaptation as such. This phenomenon is most likely a matter of planning culture as it was also outlined by the authors of the respective country reports (Wanczura, 2006, 2010; Sapountzaki & Wassenhoven, 2010; Saurí, 2010).

> *Hypothesis 2*: The legal framework and the political-administrative system significantly determine how national adaptation responses are designed and by which institutions they are implemented.

The planning systems of the analysed countries show large differences. However, this paper underlines that spatial planning is currently only given minor attention in most adaptation strategies across Europe and is rather seen as one of many other sectors. In this context, there is no clear evidence whether planning systems, which are characterized as performing planning or those whose instruments underline their conforming character, are better equipped for adaptation to climate change. It is, according to this study, rather a matter of political willingness and capacity building than particular instruments, planning has at hand. The UK and the Netherlands can be seen as countries which are in front of performing the role of spatial planning for adaptation, although they represent completely different planning cultures—performing and conforming planning. What they share is the belief that adaptation as a cross-cutting issue has to be addressed by a cross-sectoral planning approach.

> *Hypothesis 3*: Spatial planning has the potential to play an important role in climate change adaptation due to its integrative, cross-sectoral character.

As it was outlined in Section 4, there are a variety of potential contributions of spatial planning to both, coping with rapid-onset hazards as well as impacts of creeping changes in climate stimuli and the improvement of adaptive capacities. What is widely missing is a comprehensive approach which makes use of this variety of opportunities. The current situation in Europe is more like a puzzle: the different countries developed and partly implemented single pieces of this puzzle—almost independently from each other. At this stage, the Territorial Agenda as a guiding policy document comes into play which is addressed to all Member States. Thus, it is time to emphasize the cross-cutting nature of adaptation and to figure out the potential richness of planning approaches to climate change. However, there is also the issue of capacity building within the planning systems. As the UK Audit Committee emphasized (House of Commons, 2010), issuing planning guidance is not enough to ensure change. A revised Territorial Agenda has to be supported by improved skills and capacity of planning departments in order to improve decision-making. This should also be emphasized by the Territorial Agenda itself.

Note

1. This paper is based on results of a study on behalf of the German Federal Ministry for Transport, Building and Urban Development (Greiving & Fleischhauer, 2010).

References

Angignard, M. (2010) French country report. Working paper, Dortmund. Available at http://www.bbsr.bund.de/cln_016/nn_139382/BBSR/DE/FP/MORO/Studien/2009/NatStrategienKlimawandel/05__Veroeffentlichungen.html (accessed 19 January 2011).

ARGEBAU—Bauministerkonferenz (2008) *Klimaschutz in den Bereichen Bauen, Wohnen und Stadtentwicklung* (Berlin: Vorlage des Ausschusses für Stadtentwicklung, Bau- und Wohnungswesen der Bauministerkonferenz).

BMVBS—Bundesministerium für Verkehr, Bau und Stadentwicklung (2007a) *Territorial Agenda of the European Union—Towards a More Competitive and Sustainable Europe of Diverse Regions*, (Berlin), Available at http://www.bmvbs.de/cae/servlet/contentblob/44188/publicationFile/990/territorial-agenda-of-the-european-union-agreed-on-25-may-2007-accessible.pdf (accessed 19 January 2011).

BMVBS—Bundesministerium für Verkehr, Bau und Stadentwicklung (2007b) *The Territorial State and Perspectives of the European Union*, (Berlin), Available at http://www.bmvbs.de/cae/servlet/contentblob/44190/publicationFile/991/the-territorial-state-and-perspectives-of-the-european-union.pdf (accessed 19 January 2011).

Brooks, M., Gagnon-Lebrun, F., Harvey, H. & Sauvé, C. (2009) *Prioritizing Climate Change Risks and Actions on Adaptation—A Review of Selected Institutions, Tools, and Approaches*, Report by the Government of Canada, Ottawa: Policy Research Initiative.

CIRCLE ERA-Net (2009) Outcomes of the 1st International CIRCLE Workshop on Climate Change Adaptation, Budapest: Hungarian Meteorological Service.

Cools, M., Fürst, D. & Gnest, H. (2003) *Parametrische Steuerung—Operationalisierte Zielvorgaben als neuer Steuerungsmodus in der Raumplanung* (Berlin: Peter Lang).

Davoudi, S. (2010) The national country reports: United Kingdom. Working paper, Newcastle. Available at http://www.bbsr.bund.de/cln_016/nn_139382/BBSR/DE/FP/MORO/Studien/2009/NatStrategienKlimawandel/05__Veroeffentlichungen.html (accessed 19 January 2011).

DEFRA—Department for Environment, Food and Rural Affairs (2010) *National Indicator 188: Adapting to Climate Change*, (London), Available at http://archive.defra.gov.uk/corporate/about/with/localgov/indicators/ni188.htm (accessed 19 January 2011).

European Commission (2009a) *White Paper Adapting to Climate Change: Towards a European Framework for Action*, COM(2009) 147 final, Brussels. Available at http://eur-lex.europa.eu/LexUriServ/LexUriServ.do?uri=COM:2009:0147:FIN:EN:PDF (accessed 19 January 2011).

European Commission (2009b) *Staff Working Document Accompanying the White Paper Adapting to Climate Change: Towards a European Framework for Action IMPACT Assessment*, (Brussels), Available at http://eur-lex.europa.eu/LexUriServ/LexUriServ.do?uri=SEC:2009:0388:FIN:EN:PDF (accessed 19 January 2011).

Fleischhauer, M. (2010) The Netherlands—Country report. Working paper, Dortmund. Available at http://www.bbsr.bund.de/cln_016/nn_139382/BBSR/DE/FP/MORO/Studien/2009/NatStrategienKlimawandel/05__Veroeffentlichungen.html (accessed 19 January 2011).

Fleischhauer, M., Greiving, S. & Wanczura, S. (2006) *Natural Hazards and Spatial Planning in Europe*, pp. 37–54 (Dortmund: Dortmunder Vertrieb für Bau- und Planungsliteratur).

Friend, J. K. & Jessop, W. N. (1973) *Entscheidungsstrategie in Stadtplanung und Verwaltung*, Bauwelt Fundamente, Vol. 36, pp. 44 ff (Düsseldorf: Bertelsmann).

Greiving, S. (1998) *Bauleitplanung zwischen Rechtsstaatlichkeit und Praktikabilität* (Dortmund: Dortmunder Vertrieb für Bau- und Planungsliteratur).

Greiving, S. & Fleischhauer, M. (2010) National strategies of European countries for climate change adaptation: A review from a spatial planning and territorial development perspective. BMVBS-Online-Publikation 21/10. Available at http://www.bbsr.bund.de/cln_016/nn_762752/BBSR/EN/Publications/BMVBS/Online/2010/ON212010.html (accessed 19 January 2011).

Greiving, S., Fleischhauer, M., Lindner, C., Lückenkötter, J., Peltonen, L., Juhola, S., Niemi, P., Vehmas, J., Davoudi, S., Achino, E., Langeland, O., Langset, B., Medby, P., Sauri, D., Martín-Vide, J., Olcina, J., Padilla, E., Vera, F., Holsten, A., Backman, B., Schmidt-Thomé, P., Jarva, J., Tarvainen, T., Kruse, S., Schneller, K., Csete, M., Chicos, A. & Tesliar, J. (2010) *ESPON CLIMATE—Climate Change and Territorial Effects on Regions and Local Economies*, Revised Interim Report, Applied Research Project 2013/1/4, Dortmund.

Her Majesties Stationary Office (Ed.) (2008). Climate Change Act 2008. Available at http://www.legislation.gov.uk/ukpga/2008/27/pdfs/ukpga_20080027_en.pdf (accessed 19 January 2011).

House of Commons (2010) *Adapting to Climate Change. Sixth Report of Session 2009–10.*, London, Available at http://www.publications.parliament.uk/pa/cm200910/cmselect/cmenvaud/113/113.pdf (accessed 19 January 2011).

IPCC—Intergovernmental Panel on Climate Change (2007) *Climate Change 2007: Synthesis Report—An Assessment of the Intergovernmental Panel on Climate Change.*, Geneva, Available at http://www.ipcc.ch/pdf/assessment-report/ar4/syr/ar4_syr.pdf (accessed 19 January 2011).

Johnson, D. B. (1991) *Public Choice—An Introduction to the New Political Economy* (Mountain View, CA: Bristlecone Books).

Kabat, P., Van Vierssen, W., Veraart, J., Vellinga, P. & Aerts, J. (2005) Climate proofing the Netherlands, *Nature*, 438(7066), pp. 283–284.

Karaczun, Z. M., Kassenberg, A. & Sobolewski, M. (2009) *Polityka klimatyczna Polski—Wyzwanie XXI wieku* (Warszawa: Polski Klub Ekologiczny Okreg Mazowiecki).

Larsson, G. (2006) *Spatial Planning Systems in Western Europe. An Overview* (Delft: IOS Press).

LRAP (2010) *Adapting to Climate Change. Guidance Notes for NI188,* Version 1.8., London, Available at http://archive.defra.gov.uk/corporate/about/with/localgov/indicators/documents/ni188-guidance.pdf. (accessed 19 January 2011).

Massey, E. & Bergsma, E. (2008) *Assessing Adaptation in 29 European Countries*, Report W-08/20, Institute for Environmental Studies (Amsterdam: Vrije Universiteit).

Meister, H.-P., Kröger, I., Richwien, M., Rickerson, W. & Laurent, C. (2009) *Schwimmende Häuser und Moskitonetze: Weltweite Strategien zur Anpassung an den Klimawandel* (Bensheim: Nationale Strategien und Projektbeispiele).

Mickwitz, P., Aix, F., Beck, S., Carss, D., Ferrand, N., Görg, C., Jensen, A., Kivimaa, P., Kuhlicke, C., Kuindersma, W., Máñez, M., Melanen, M., Monni, S., Pedersen, A. B., Reinert, H. & Van Bommel, S. (2009) *Climate Policy Integration, Coherence and Governance.* PEER Report No. 2, Helsinki. Available at http://files.ifok.de/all/klima/IFOK-Studie_Klimaanpassung.pdf(accessed 19 January 2011).

Neary, P. J. (2001) Of hype and hyperbolas: Introducing the new economic geography, *Journal of Economic Literature*, 39(2), pp. 536–561.

Ribeiro, M., Losenno, C., Dworak, T., Massey, E., Swart, R., Benzie, M. & Laaser, C. (2009) *Design of Guidelines for the Elaboration of Regional Climate Change Adaptations Strategies* (Vienna: Ecologic Institute).

Rivolin, U. J. (2008) Conforming and performing planning systems in Europe: An unbearable cohabitation, *Planning, Practice and Research*, 23(2), pp. 167–186.

Roggema, R. (2009) *Adaptation to Climate Change: A Spatial Challenge*, pp. 1–58 (Dordrecht: Springer).

Sapountzaki, P. & Wassenhoven, L. (2010) *Adaptation to Climate Change/Greek Case Study.* Athens. Available at http://www.bbsr.bund.de/cln_016/nn_139382/BBSR/DE/FP/MORO/Studien/2009/NatStrategienKlimawandel/05__Veroeffentlichungen.html (accessed 19 January 2011).

Saurí, D. (2010) The territorial dimension of adaptation to climate change in Europe: The case of Spain. Working paper, Barcelona. Available at http://www.bbsr.bund.de/cln_016/nn_139382/BBSR/DE/FP/MORO/Studien/2009/NatStrategienKlimawandel/05__Veroeffentlichungen.html (accessed 19 January 2011).

Schröter, D., Acosta-Michlik, L., Arnell, A. W., Araújo, M. B., Badeck, F., Bakker, M., Bondeau, A., Bugmann, H., Carter, T., De La Vega-Leinert, A. C., Erhard, M., Espiñeira, G. Z., Ewert, F., Fritsch, U., Friedlingstein, P., Glendining, M., Gracia, C. A., Hickler, T., House, J., Hulme, M., Kankaanpää, S., Klein, R. J. T., Krukenberg, B., Lavorel, S., Leemans, R., Lindner, M., Liski, J., Metzger, M. J., Meyer, J., Mitchell, T., Mohren, F., Morales, P., Moreno, J. M., Reginster, I., Reidsma, P., Rounsevell, M., Pla, E., Pluimers, J., Prentice, I. C., Pussinen, A., Sánchez, A., Sabaté, S., Sitch, S., Smith, B., Smith, J., Smith, P., Sykes, M. T., Thonicke, K., Thuiller, W., Tuck, G., Van Der Werf, G., Vayreda, I., Wattenbach, M., Wilson, D. F., Woodward, F. I., Zaehle, S., Zierl, R., Zudin, S. & Cramer, W. (2004) *ATEAM Final Report. Section 5 and 6 and Annex 1 to 6*, Detailed report, related to overall project duration, Potsdam.

Schuster, P. (2008) *Klimaanpassungsstrategien in europäischen Nachbarländern—ein Werkstattbericht.* Anpassungsstrategien an den Klimawandel in EU27 und ihre potentielle Bedeutung für die Raumentwicklung in Deutschland, BBR.

Stern, M. (2006) *The Economics of Climate Change—The Stern Review* (Cambridge: Cambridge University Press).

Swart, R., Biesbroek, G. R., Binnerup, S., Carter, T., Cowan, C., Henrichs, T., Loquen, S., Mela, H., Morecroft, M., Reese, M. & Rey, D. (2009) *Europe Adapts to Climate Change. Comparing National Adaptation Strategies*, PEER Report No. 1, Helsinki.

Vajdovich, E. (2010) Hungarian country report. Working paper, Budapest. Available at http://www.bbsr.bund.de/cln_016/nn_139382/BBSR/DE/FP/MORO/Studien/2009/NatStrategienKlimawandel/05__Veroeffentlichungen.html (accessed 19 January 2011).

Wanczura, S. (2006) Poland, in: M. Fleischhauer, S. Greiving & S. Wanzura (Eds) *Natural Hazards and Spatial Planning in Europe*, pp. 127–144 (Dortmund: Dortmunder Vertrieb für Bau-und Planungsliteratur).

Wanczura, S. (2010) *Polish Country Report*. Dortmund. Available at http://www.bbsr.bund.de/cln_016/nn_139382/BBSR/DE/FP/MORO/Studien/2009/NatStrategienKlimawandel/05__Veroeffentlichungen.html (accessed 19 January 2011).

Wassenhoven, L., Dousi, E., Housianakou, M., Koutalakis, H. & Lalos, A (2009) *Climate Change, Environmental Vulnerability and European Challenges*, ISTAME (Institute of Strategic and Development Studies), Green Development, Reports for Consultation, Athens, pp. 3–54 (in Greek).

World Bank (2008) *The Global Monitoring Report 2008.*, Washington, Available at http://siteresources.worldbank.org/DEVCOMMINT/Documentation/21716432/DC2008-0003(E)GMR2008.pdf (accessed 19 January 2011).

Climate Risk and Security: New Meanings of "the Environment" in the English Planning System

SIMIN DAVOUDI*,**

*School of Architecture, Planning and Landscape, Newcastle University, Newcastle upon Tyne, UK,
**Newcastle Institute for Research on Sustainability (NIReS), Newcastle University, Newcastle upon Tyne, UK

ABSTRACT *Planning in England*[1] *represents an important arena for the development and contestation of environmental discourses. Over the last century the changing assumptions about human-nature relationship have led to numerous meanings of "the environment" in planning. These have in turn influenced the choices made between: preserving, enhancing, protecting, compromising, trading, exploiting or guarding against, the environment. While recognizing the nuances of the environmental discourses, this paper identifies eight distinct meanings of the environment in contemporary plans including the environment: as local amenity, as heritage landscape, as nature reserve, as storehouse of resources, as tradable commodity, as problem, as sustainability and as risk. The latter has emerged as a result of growing climate change awareness. The paper argues that, while the emphasis on climate change mitigation has reinforced some aspects of the sustainability discourse, the adaptation agenda has introduced a new meaning of the environment as risk. This portrays the environment not so much in terms of assets to be sustained for human benefit, but in terms of threats against which human well-being should be safeguarded. Framed in the language of risk and security, this new discourse is bringing to the fore some of the outmoded approaches to planning.*

1. Introduction

"The potential for us to make progress with environmental issues is limited by the basic assumptions that we make about nature, the unspoken, often unrecognized perspective from which we view our environment" (Botkin, 1990).

This paper concurs with the above in suggesting that there is no single overarching meaning of "the environment"; that the way in which environment is defined has evolved with our changing assumptions about nature and our relation to it. Nature in this context is understood neither as the essential quality of something (*natura rerum* or nature of things) nor as an inherent force which drives the world but, following Marshall (1994), as the material world as a whole.

The changing meanings of the environment are both reflected in and constituted through environmental discourses[2] in planning. The institutions and practices of planning provide key arenas in which environmental discourses are developed, mobilized and contested. Like all forms of narratives, environmental narratives in planning are selective abstractions which amplify one meaning of the environment and marginalize others. Their formation is a contested political act, infused with power relations. As Foucault suggests, their power "derives not so much from the abstract ideas they represent as from their material basis in the institutions and practices that make up the micro-political realm" (quoted in Barnes & Duncan, 1992, p. 9). The outcome of this dynamic and contested process is not merely symbolic. It also has material consequences for the choices made in planning practice about preserving, enhancing, protecting, compromising, trading, exploiting or guarding against the environment. It determines how much weight should be given to the environment compared with social and economic concerns. It shapes planners' conception of the "sustainable city" and how it might be achieved. The formation of climate change discourses with their emphasis on risk and security can have a profound effect on the treatment of the environment in planning. They can shift the emphasis from what we do to nature towards what nature does to us.

The aim of this paper is to unravel different meanings of "the environment" in the English planning system in the context of changing assumptions about human-nature relationships. Particular emphasis is put on emerging climate change discourses and the way in which they are shaping perceptions of sustainability. The analyses are based on a review of relevant European (such as European Spatial Development Perspective (ESDP) and its successors) and national policies (particularly related to climate change) as well as a content analysis of 28 planning documents produced in England in the 2000s (listed in Table 1). The latter include the Regional Spatial Strategies (RSS) for all nine English regions including London. The RSSs were given statutory weight in 2004 and as a result could exert more influence on the environmental narratives of the local plans, particularly in relation to climate change. Based on these analyses as well as the review of the existing literature, the paper identifies eight distinct meanings of the environment in planning. These do not neatly succeed each other. Instead, the "new" perspectives have often been added to the "old" ones creating a multi-layered, although not necessarily complementary, narrative of the environment in contemporary plans. It is argued that the onset of climate change mitigation and adaptation have, on the one hand, brought some of the "old" environmental discourses to the foreground, and on the other hand, introduced an entirely new meaning of the environment in planning; one which portrays the environment as a risk and a threat to human well-being. This in turn is renewing a number of outmoded practices in planning. The paper is structured under three main sections. After this introduction, section two provides an account of the eight environmental discourses in planning, and section three concludes the paper.

2. Meanings of "The Environment" in Planning

"There is a great variety of distinct and often radically different (if not antagonistic) natures" (Swyngedouw, 2007, p. 36).

Table 1. Sample of the planning documents produced in England in the 2000s

Title	Publication date
National planning policy statements	
PPS7: Sustainable Development in Rural Areas	2004
PPS22: Renewable Energy	2004
PPS9: Biodiversity and Geological Conservation	2005
PPS1: Delivering Sustainable Development	2005
Supplement to PPS1: Planning and Climate Change	2007
Supplement to PPS1: Eco-towns	2009
PPS4: Planning for Sustainable Economic Growth	2009
PPS3: Housing	2010
PPS25: Development and Flood Risk	2006, revised 2010
Supplement to PPS25: Development and Coastal Change	2010
Regional spatial strategies	
Regional Spatial Strategy for the East Midlands (RSS8)	March 2005
A Clear Vision for the South East: The South East Plan Core Document (Draft for Submission to Government)	March 2006
Draft Regional Spatial Strategy for the South West 2006–2016	No date
Regional Spatial Strategy for the West Midlands	January 2008
East of England Plan: The Revision to the East of England Spatial Strategy	May 2008
The Yorkshire and Humberside Plan, Regional Spatial Strategy to 2026	May 2008
The North East of England Plan—Regional Spatial Strategy to 2021	July 2008
North West of England Plan—Spatial Strategy to 2021	September 2008
The London Plan: Spatial Development Strategy for Greater London	October 2009
Local development plan documents	
Core Strategy and Development Control Policies. Chelmsford Borough Council Local Development Framework 2001–2021 (Adopted)	February 2008
(City of Lancaster) Local Development Framework Core Strategy 2003–2021 (Adopted)	July 2008
London Borough of Richmond upon Thames: Local Development Framework Core Strategy (Adopted)	April 2009
(Lake District National Park Authority) Local Development Framework Core Strategy (including Proposals Map)—(Adopted)	October 2010
East Riding of Yorkshire: Preferred Approach Core Strategy	May 2010
London Borough of Merton: Local Development Framework Core Strategy 2010 Pre-submission Representations Publication	August–September 2010
Climate change strategies	
Rising to the Challenge: A Climate Change Action Plan for England's North West 2010–2012	No date
(The East of England Climate Change Partnership) Climate Change Action Plan—Draft	2009
The Draft Climate Change Adaptation Strategy for London (Public Consultation Draft)	February 2010

Concerns about the environment have a long history in planning. The rise of the planning movement in the late nineteenth century in England and elsewhere in the UK as a response to urban industrialization was both socially and environmentally motivated. One of its key characteristics was a desire to halt or slow down the unprecedented level of urbanization which was encroaching into the countryside and degrading nature.

Since then environmental concern has been a major preoccupation in planning. However, the meanings given and the values attached to "the environment" have changed substantially over time. An early attempt to trace this evolution was made by Howard Newby who argued that, "environmental concerns in the United Kingdom mark a shift in perspective from an amenity-led to an ecology-led approach to environment" (Newby, 1990, p. 3). Later, Whatmore and Boucher (1993) explored the role of planning discourses and practices in the social construction of nature and its manifestation in the spatial tensions between development and conservation, arguing that until the 1980s the conservation narrative was dominant. They suggested that planning had played an important part in regulating the spatial contours of the "ontological divide between society and nature", and that such a divide is represented in the persistent dualism between "built" and "natural" environments (Whatmore & Boucher, 1993, p. 176). Building on these analyses, Healey and Shaw (1994, p. 427) provided a more detailed history of environmental discourses in planning and identified five strands including: welfarist-utilitarianism (1940s/1950s), growth management (1960s), active environmental care (1970s), marketized utilitarianism (1980s) and sustainable development (1990s).

Based on a content analysis of contemporary English planning documents, this paper argues that firstly, the differentiation of environmental discourses in planning is more nuanced than previously articulated. Secondly, since the late 1990s climate change awareness has introduced an entirely new discourse. It is therefore possible to identify eight distinct meanings of the environment in planning, including the environment: as local amenity, as heritage landscape, as nature reserve, as storehouse of resources, as tradable commodity, as problem, as sustainability and as risk. In the following section the key features and the origin of these discourses are discussed, and their historical and contemporary leverage in planning are examined. Examples from past planning documents (using primary and secondary sources) are drawn upon to illustrate the points made.

2.1 *Environment as Local Amenity*

Seeing the environment as local amenity is one of the oldest and most established utilitarian conceptions of the environment in planning. In fact it predates the use of the term "environment". Intangible amenities such as aesthetic and recreational values have provided the rationale for conservation narratives in planning since its inception. Their origin can be traced back to the Romantic period of the nineteenth century and its reaction to the dominant "mechanical" view of the world. For Romantics, nature had symbolic values and was a source of delight and inspiration. "Nature was where industry was not" (Williams, 1972, p. 159 quoted in MacNaghten & Urry, 1998, p. 13); it was the wilderness in which one could take refuge in the face of the accelerating pace of urban industrialization. This green romanticism underpinned some of the anti-urban reactions of the growing environmental campaigns in the mid-nineteenth century (Lowe and Goyder, 1983). It was also an influential driver of the planning movement and the emergence of two markedly different attitudes towards the countryside in the early twentieth century. One represented the countryside as a source of inspiration and a powerful "magnet" whose amenities should be woven into the urban fabric to improve the standard of city living. The other represented the countryside as a nostalgic rural life to be defended against urban sprawl. The former was epitomized in the Garden City movement and Ebenezer Howard's attempt to combine the best of town with the best of country. The latter was manifested in the reactions of his opponents whose idyllic view of the countryside was represented in metaphors similar to the classic Greek and Roman images of "mother nature". For example, Thomas Sharp, a leading planning figure, was insisting on a clear separation of the "masculine" urban from the "feminine" rural, claiming that,

"Tradition has broken down ... Urban influences neutralize the country .... The strong, masculine virility of the town; the softer beauty, the richness, the fruitfulness of that mother of men, the countryside, will be debased into one sterile, hermaphrodite beastliness" (Sharp, 1932, p. 11).

Although the Garden City idea has remained as a source of inspiration, it was the dichotomous view of urban versus rural which became the established discourse in the early twentieth century planning system. It found powerful advocates among some of its most celebrated figures including Patrick Abercrombie who was of the view that although the 1932 Planning Act had included "statutory powers to deal with both" town and country, there "should be no attempt at fusion between the two ... town should be town and country should be country; urban and rural can never be interchangeable adjectives" (Abercrombie, 1933, p. 177). Therefore, his 1944 Greater London Plan established one of the most enduring planning orthodoxies, the urban containment which became institutionalized through the Green Belt.

Throughout the 1940s and 1950s, seeing the environment as local amenity and a backcloth to development remained the dominant perspective (Healey & Shaw, 1994). The goal of planning was to ensure that the natural beauty and the pleasing landscape are utilized in new development projects for the visual and recreational benefits of humans. This view of the environment continued in the subsequent planning documents in statements such as: "the function of (landscape) framework varies ... Sometimes it will act as screen, sometimes as a backcloth" (LCC, 1974, p. 128), or "the need to protect the intrinsic environment and visual qualities of the countryside" (LCC, 1994, p. 17). Seeing the environment as a setting for development has retained its discursive leverage in contemporary plans. For example, the local plan for Merton (LBM, 2010, p. 4) refers to "the wide open spaces" as "beauty spots" which are "important ... (for) everyone's well-being". Chelmsford local plan is "recognising the important visual and landscape function that they (Green Wedges) have for the town" (CBC, 2008, p. 31).

2.2 *Environment as Heritage Landscape*

Like amenity, heritage is a nebulous yet highly venerated concept. It refers to both built and natural environment but is particularly associated with landscape. Like amenity it has a long history dating back to the mid-nineteenth century. However, unlike amenity which began as a welfare-oriented discourse, the early approaches to heritage were highly elitist. As Newby (1990, p. 6) argues, "Environmental artefacts were to be preserved "*for* 'the Nation', but *from* 'the public' which was frequently regarded as unappreciative and philistine, representing a threat to this 'national heritage'". These sentiments shared some of the values advocated by the green romantics in terms of putting the emphasis on preservation of a pre-industrial past and protection of the "national" heritage from urban industrialization. During the interwar period this elitist preservationism came into conflict with the demand for wider access to the countryside. The tension centred on the perceived conflicts between the agricultural production value of land and its recreational, consumption value. The former was upheld by the farmers—who are traditionally seen as stewards of the countryside- and the environmental groups such as the Council for Preservation of Rural England.[3] The latter was supported by the Labour Party, the Unions and the campaign groups such as the Ramblers' Association (Marsden *et al.*, 1993; Healey & Shaw, 1994).

An important dimension of seeing the environment as heritage landscape, as opposed to local amenity, is the assignment of hierarchical values to it ranging from national to local significance and resulting in their descending level of protection. At the top of the hierarchy are National Parks whose statuary designation in 1949 was a political victory for the alliance in favour of public rights of access to the countryside. Further down the hierarchy are the Areas of Outstanding Natural Beauty (AONB), and lower still are various sites of locally significant landscapes. The precursor to this ordering of nature into hierarchical categories is the eighteenth century's burgeoning biological classifications of all "living things". By creating mental pigeon-holes for rational and comprehensive arrangement of all species, natural scientists aimed to make them amenable to imposition of order, not by the Creator as was previously believed, but by human mind (Bowler, 1992). As Foucault (quoted in Bowler, 1992, p. 141) suggests, this was the beginning of the modernist passion for order, fixity and control; a passion upon which planning as a modernist project was founded.

The Location of designated landscapes on the hierarchy results in different degrees of protection assigned to them. So, those at the bottom become more vulnerable to development pressures than those at the top. This, however, does not suggest that sites at the top of the ladder are immune from development. For example mineral (coal, aggregates, metal, etc.) extraction can take place even in National Parks as long as it is perceived to be in the "national interest". Hence, mineral operations are a source of tension not only between development and conservation, but also between two meanings of the environment: the environment as storehouse of resources (discussed below) and the environment as heritage landscape. The conception of the environment as heritage landscapes that are ranked in order of their perceived significance still occupies a central place in planning discourses. Every contemporary plan has a section with a list of designated sites that are often presented as "environmental capital" of which "the community can be justifiably proud" (LCC, 2008, p. 55).

2.3 *Environment as Nature Reserve*

Although there may be similarities between the environment as heritage landscape and as nature reserve, there is a fundamental difference between the two. The latter is the only conception of the environment in planning which is not rooted in the anthropocentric view of nature. The anthropocentric view places humans at the centre of Universe and considers nature to be at their service. An early and striking example of this view is Francis Bacon's assertion that: "Man ... may be regarded as the centre of the world... For the whole world works together in the service of man; and there is nothing from which he does not derive use and fruit...insomuch that all things seem to be going about man's business and not their own" (Quoted in Marshall, 1994, p. 184). By contrast, the biocentric view of nature considers humans as an integral part of nature and as members of an interconnected "web of life" (Marshall, 1994). It therefore challenges the 1970s discourse of environmental stewardship in planning which is presented in Healey and

Shaw (1994) and considers humans as part of nature rather than its steward or its master. Although the bio-centric view of human-nature relationship predates the anthropocentric perspective, its influence has been sidelined after the scientific revolution of the seventeenth century. Since then it has remained subservient to the dominant anthropocentric view and has only recently re-emerged in the form of radical ecology.

Seeing the environment as nature reserve and showing concerns not just for human benefits, but also for non-human species is the closest that planning discourses have come to a bio-centric worldview. However, even this has often retained a hint of functional utilitarianism. Although early signs of a bio-centric perspective can be found in Patrick Geddes' idea of the "natural region", its real influence on planning discourses began in the 1970s and through a series of international designations. An early example is the designation of Ramsar Sites which are wetland areas protected essentially for the well-being of migrating birds and not human enjoyment. More recent examples include various European designations such as the Special Areas of Conservation and the Special Protection Areas, which are safeguarded for their biodiversity and habitats values. As with landscape sites, embedded in these designations is a hierarchical order which reflects "the specific importance of particular habitats" (CBC, 2008, p. 108). Thus, Local Wildlife Sites are construed as having less value than those with national or international safeguards.

The bio-centric perspective in planning saw a further boost in the 2000s through the discourses of "green corridors" and "green networks" because the main motivation for protecting and creating such "green links" is "to create safe species movement and havens for nature" (LBM, 2010, p. 130). An important feature of these new concepts is their articulation of nature as relational and fluid rather than bounded and fixed. Although valuing and protecting nature for nature's sake has not been a prominent environmental discourse in planning, it has continued its somewhat marginal presence. The RSS for East of England, for example, urges the planning authorities to "seek to ... protect, for their own sake, all important aspects of the countryside" (GOEE, 2008, p. 50).

2.4 *Environment as Storehouse*

Seeing the environment as a storehouse of resources and functions that can be exploited at will for human benefits has been a powerful and enduring conception of the environment in industrial societies (Verhagen, 2008) with significant influence on environmental discourses in planning. It is rooted in the seventeenth century scientific revolution and particularly the rise of Newtonian mechanics[4] which marked a shift in the understanding of human relation to nature. Nature began to be perceived as a machine, yet designed and maintained by the divine Creator. By the eighteenth century the divinity of its origin was gradually replaced by scientific explanation. The Universe was compared with a "Great Wheel" which was "kept in circulation" by its own internal qualities (quoted in Botkin, 1990, p. 106). Nature was seen as gigantic clockwork whose wheels and springs were operating according to some hidden rules and structures. Hence, scholars, such as Francis Bacon, were of the view that by uncovering the secrets of nature through rationalist search human race could control the material world (Hollis, 2003). This rational view, presented ironically in a gendered and romantic language, claimed that, "science would "strip the veil" from nature so that she revealed her innermost self" (quoted in Bowler, 1992, p. 91).

Reducing nature to a material system gave humans both the means and the right to exploit it. Detached from its former symbolic values nature was no longer cherished as a source of inspiration. As Bowler (1992, p. 89) argues, the mechanistic view of nature was necessary for legitimization of its exploitation. Thus, the age of discovery and scientific revolution was driven by a desire not just to explore, but also to exploit nature (Bowler, 1992); a desire that was materialized on an industrial scale in the nineteenth century. This was the time when environment began to be seen as a storehouse of goods, resources and functions, all of which were under human control and for human exploitation. Adam Smith, the founder of modern economics, for example, considered the environment as "a storehouse of raw materials for man's ingenuity" (quoted in Worster, 1994, p. 53). Conquering nature through technological advances and utilizing environmental resources for economic growth are still considered as hallmarks of progress in industrial societies. In the English planning system, however, this extreme view of the environment as a mere input into industrial processes has never achieved a noticeable leverage. This is partly because planning itself emerged as a rational instrumental response to the excesses of environmental exploitations which was associated with urban industrialization in the late nineteenth century. Nevertheless, a more subtle version of this extreme view has had a profound influence on planning discourses. This is reflected in the utilitarian approaches to the environment and its treatment as a container of, material resources, scientific repository and functional services, and also as a "sink" for processing waste and pollution. These will be elaborated in turn.

2.4.1 *Container of material resources.* Seeing the environment as a storehouse with unlimited material resources was a particularly influential view during the 1960s and 1970s (Healey & Shaw, 1994). Paradoxically, this was a period when concerns about scarcity of environmental resources and fragility of nature were also on the rise (see Section 2.6). However, these concerns hardly influenced planning discourses until the late 1980s. Even then they were treated as secondary to economic priorities (Davoudi *et al.*, 1996). A key material resource whose extraction has been at the centre of development versus conservation conflict is minerals, as mentioned above. A typical narrative in planning documents which reveals this tension can be found in the Lancashire Structure Plan (LCC 1994, p. 38). It states that although "the countryside will be protected for its own sake ... the requirement of minerals industry must be addressed". A frequently cited justification for the exploitation of environmental resources is that it is in "public interest" where "public" is often equated with nationally determined policy goals. Often potential "local environmental" losses are justified by perceived "national economic" gains (Davoudi, 1997).

2.4.2 *Scientific repository.* Treating the environment as a scientific repository to be explored for human benefits is another way of seeing the environment as storehouse. In the planning documents, it is most explicitly represented through the designated Sites of Special Scientific Importance (SSSI). Like other types of designations, these portray the environment as static and bounded. Nature is neatly and hierarchically classified into parcels of land with clear delineation. It is then "sieved" through a mapping exercise which eliminates the higher order sites (considered as areas of "constraint") from consideration for future development. The parcelling of nature along with the "sieve map" process make the complex environmental relationships amenable to planning control. These demarcations are part of "planning technologies" for determining which bits of the environment are to be protected and which bits are to be lost to development.

2.4.3 *Functional services.* Conceiving of the environment as a container of functional services often coincides with the amenity discourse in planning. Examples of the functional utilitarian view of the environment can be found throughout the history of planning (Healey & Shaw, 1994; Davoudi *et al.*, 1996) and are still abundant in contemporary plans. It is most frequently represented in the approaches to open spaces which are seen as "green lungs" for the built up areas. For example, the local plan for Chelmsford (CBC, 2008, p. 56) states that, "the importance of open spaces relates to their function and also amenity value ... in providing a "green lung" and visual break in the built environment". Since the 2000s, climate change adaptation and mitigation agenda has reinforced this functional view. One example is seeing the environment as a source of renewable energies which in a national planning statement (DCLG, 2007, p. 6) is described as flowing repeatedly from: "the wind, the fall of water, the movement of the oceans, from the sun and also from biomass". Such statements combine a twenty-first century functional view of the environment with a nineteenth century's lyrical language of green romanticism.

This functional approach is more explicitly advocated by climate adaptation discourses. These have reinvigorated the perception of the environment as a container of eco-system goods and services. Increasingly, emphasis is put on multi-functionality of environmental services, implicitly implying some kind of utilitarian efficiency. Green spaces in particular have been at the centre of attention. Added to their traditional function as "breathing lungs" for built up areas are a multitude of new functions such as: cooling ambient temperature, encouraging biodiversity, draining flood waters, sequestrating carbon, providing food security, being an outdoor classroom and maintaining human health. The latter refers not simply to the recreational and leisure functions of the environment, as in the 1940s and 1950s, but also to its cardiovascular and fitness function. The following extract from the London adaptation plan (GLA, 2010, p. 6) exemplifies this language of multi-functionality.

"Becoming a greener, leafier city is more aesthetically pleasing, adds to a sense of well-being and reinforces London's position as one of the best big cities in the world. At the same time, urban green space reduces flood risk and cools the city in hot weather. Alongside the eco-creativity required to weatherproof our city also comes considerable untapped employment opportunities".

Two points are worth mentioning. Firstly, the inclusion of the old and new functional values into the capacious discourse of multi-functionality does not necessarily lead to a "grand unification" (Hebbert, 2009, p. 365) of their different utilities which inevitably remain in competition for pre-eminence, or even in conflict, in planning discourses. Secondly, while the climate adaptation discourse of "eco-system goods and services" is ultimately a functional utilitarian one, it does represent a tentative, yet significant change of attitudes because, it puts the emphasis on making space for nature (e.g. freeing up land for flooding or coastal retreat) rather than on controlling nature (e.g. building flood and coastal defences); it puts "Nature first" (Gibbs *et al.*, 2007).

2.4.4 *Sink for waste and pollution.* One of the functional utilities associated with, though not always explicitly related to, the environment in planning documents is the "sink" function. Treating the environment as a giant sink which assimilates waste and pollution is the flip side of the industrial image of nature as storehouse. Considering pollution as an inevitable by product of economic growth and progress has been a key feature of modernism. The idea is that although attempts should be made to reduce pollution, these should not be at the cost of slowing down progress. The initial planning response to this challenge was to keep the polluting industries intact but separate them from residential areas through zoning. As with the amenity, the emphasis was on local pollutions, such as noise, dirt, smell and smoke that were known through direct experiences and their impacts on local quality of life (Healey & Shaw, 1994). The wider implications were either unknown or neglected and therefore did not enter into planning discourses until the 1970s. Today, treating the environment as a sink can be found in a number of areas such as: the emission of greenhouse gases (GHG) to the atmosphere, the discharge of waste water to the sea, and the burying of solid waste under the ground. The latter, which has been the most explicit manifestation of this treatment of the environment, is still the dominant practice in the UK. The term "sink" itself, which was previously hard to find in planning documents because of its negative connotation, is now appearing frequently. For example, a national policy statement (DCLG, 2007, p. 5) refers to "trees and vegetations" in their capacity to act as "carbon sinks".

2.5 *Environment as Tradable Commodity*

The storehouse image of the environment and the exploitation of its resources have been largely justified in planning on the ground of being in "public interest". In the 1980s, this welfare-oriented perspective gave way to a market-led approach and the portrayal of the environment as a commodity. The change was driven by the neo-liberal climate of the time and its manifestation in planning in terms of "a presumption in favour of development" (DoE, 1985). As a result, "a vision of planning as a bargaining process" and a "vision of nature as a social product valued through the market place" was constructed (Whatmore & Boucher, 1993, p. 170). Environment was treated as a commodity which could be traded with other commodities through the bargaining processes of planning gains. The assumption was that the loss of one environmental "parcel" can be offset by gaining another as long as the bargaining process did not jeopardize development projects. This was exemplified in the Countryside Commission's argument that, "what is lost in environmental terms by the development can be more than offset by some associated benefit" (quoted in Whatmore & Boucher, 1993, p. 171).

This treatment of the environment as a tradable asset continued in the 1990s and underpinned the economically driven approaches to sustainable development. Its most notable representation was the notion of "constant capital stock" coined by David Pearce and his team. In their *Blueprint for a Green Economy* (Pearce *et al.*, 1989) they suggested that for a society to be sustainable an undiminished per capita stock of capital should be passed from one generation to the next. And, that "natural capital" can be substituted with human-made capital (Pearce & Barbier, 2000). While the 1980s' "commodity" discourse treated the environment as an "object" which could be "reproduced" through human action, the 1990s' "capital stock" discourse treated it as an "object" which could be "substituted" with manufactured objects (Hamdouch & Zuindeau, 2010). The discourse of the environment as a "stock of capital" or a trading commodity continues to run through contemporary plans. For example, the local plan for Lancaster lists the city's "natural and built 'Environmental Capital'" and presents them as "a major economic asset" (LCC, 2008, p. 55). The RSS for the North West Region considers "Access to greenspace" as having "a central role to play in securing successful and sustainable economic regeneration" (GONW, 2008, p. 89). In the London Plan, measures for reducing carbon emissions are framed in terms of their economic values in statements such as: "preventative and adaptive measures will generate long term savings... and should have positive impacts on property values" (GLA, 2009, p. 94).

2.6 *Environment as Problem*

Seeing the environment as a "problem" for which solution had to be found made its "dramatic leap to the top of the political agenda" in the 1960s (Dryzek, 1997, p. 21) when, for the first time, the picture of Earth taken from space symbolized its fragility, its finite resources and its limited capacity to support life and satisfy human demand. Attention began to shift from what nature does to us and what we can get out of it, to what we do to nature and the problematic consequences of our actions. This newly discovered "problem" triggered two distinct responses: one advocated radical actions to curb human demand, the other promoted problem-solving approaches, as outlined below. In the event, it was the latter which had a major leverage in planning discourses.

A key factor in the realization of the sheer scale of environmental "problem" was the publication of *Limits to Growth* (Meadows *et al.*, 1972). This propelled the discourse of "limits" into public imagination and political agenda. The study was undertaken by a team of system modellers sponsored by a group of industrialists and academics called the Club of Rome. Their message was simple yet powerful; environmental care and concern is not just desirable (as portrayed in the "stewardship" discourses of planning at that time) but also crucial for the survival of humanity. This resonated with the earlier "carrying capacity" concerns which had been long deployed by ecologists and population biologists. The basic story line of this "survivalism" discourse is that, "human demands on the carrying capacity of ecosystems threaten to explode out of control, and draconian action needs to be taken in order to curb these demands" (Dryzek, 1997, p. 34). The opening of a "hole in the ozone layer" seemed to provide the seal of approval for the Club of Rome's foresight. Survivalism, however, did not have much leverage in the environmental discourses of planning because firstly, it dealt with aggregates such as global population and resources. Secondly, it puts the emphasis not just on "thinking" globally, but also on "acting" globally in a centrally coordinated way (Dryzek, 1997, p. 43). Both of these were well beyond the scope of the locally focused planning system. Therefore, despite occasional references to limits and carrying capacity in planning documents (which has continued to date and can be found in, for example, the RSS for the North West Region (GONW, 2008, p. 28)) survivalism did not influence planning narratives in a noticeable way.

On the contrary, what left a significant mark was the problem-solving approach. As Dryzek (1997, p. 61) suggests, problem-solving recognizes "the existence of environmental problems but treats them as tractable within the basic framework of the political economy of industrial society, as belonging in a well-defined box of their own". In planning, an early symbolic representation of the "boxing" approach was to put everything that is considered as environmental issues into one chapter of the plan and colour code them green. Within this chapter the environment was further broken down (to, for example, open space, habitat, landscape heritage, waste, etc.) and different pieces were put in smaller boxes of their own. Although this became a relatively common practice in planning immediately after the introduction of sustainable development agenda (Davoudi & Layard, 2001), later it was largely abandoned.

According to Dryzek (1997, p. 61) there are three routes through which the problem-solving approach to the environment has been implicated. These are: "administrative rationalism (leave it to the experts), economic rationalism (leave it to the market), and democratic pragmatism (or, leave it to people)". The English planning system plays a major part in the first route because administrative rationalism manifests itself in the state's institutions and practices in which planning occupies a prominent position. Its space of action on environmental "problems" has been considerably enlarged since the late 1980s mainly in response to the growing number of the EU environmental directives (on waste, habitat, water, birds, etc.). Of particular significance has been the EU requirement to carry out environmental impact assessment (EIA) for major development projects. The decision of the central government to use the planning system as the administrative space in which EIA is implemented had a profound influence on the environmental discourses in planning. It was also crucial in preventing the downgrading, or even the demise, of the planning system itself at the time when the neo-liberal pressures for deregulation and rolling back of state was at its height.

Seeing the environment as a problem for which rational administrative solutions have to be sought has remained a powerful perspective in planning. Planning documents have continued to present a combination of both "concerns" for the environmental problems and "reassurances" for the ability to solve them without having to change the existing socio-economic structures. The impact assessments in particular have reinforced the perception of planning as a process of evidence-based, impartial reasoning by technical experts. The aim is to provide reassurances that the outcome of the assessment would be in the public interest even if it may ultimately be detrimental to the environment. However, despite these limitations the EU-driven procedures and assessment techniques have strengthened the environmental discourses in planning. Concerns about, for example, waste and water management which had been overlooked for many decades are now embedded in planning narratives (Davoudi, 2000a). It is also worth mentioning that the "leave it to expert" discourse is frequently challenged in the planning arenas where the tensions between this and the "leave it to people" approach to environmental problem-solving are brought to the fore.

2.7 *Environment as Sustainability*

While the language of sustainability has been around since the eighteenth century as part of forestry practices, its pre-eminence in the environmental discourses is due to the publication of the Brundtland Report (WCED, 1987) and its famous definition of sustainable development.[5] While recognizing the significant contribution of the Report to environmental debates and actions, it is worth highlighting its underlying human-centred approach to nature which is exemplified in the following statement:

"In essence, sustainable development is a process of change in which exploitation of resources, the direction of investments, the orientation of technological development and institutional change are all in harmony and enhance both current and future potential to meet needs and aspirations" (WCED, 1987, p. 46).

This also shows the Report's assumption about a positive-sum relationship between distributive justice, economic growth and environmental protection. It is this win-win rhetoric which attracted the support of a wide spectrum of actors and agencies. Furthermore, and contrary to the widely held view, the discourse of sustainable development as represented in the Brundtland Report does not have much affinity with the older discourse of limits and survival. Although it recognizes the existence of "ultimate limits", it goes on to declare that, "Growth has no set limits in terms of population or resources use beyond which lies ecological disaster". The reason for such assumptions is the over-confidence in human ingenuity and technological solution: "accumulation of knowledge and the development of technology can enhance the carrying capacity of the resources base" (WCED, 1987, p. 45); "...we have the ingenuity to change" (WCED, 1987, p. 205). This is a powerful rhetoric and one which has been fully embraced by the UK government and the planning system since the early 1990s. Its omnipresence in planning documents is reflected in numerous narratives about striking a balance between economic, social and environmental goals.

This dominant positive-sum view of sustainability is a weak form of ecological modernization (Hajer, 1995; Lafferty & Meadowcroft, 1996; Davoudi, 2000b) which draws on rational technical instruments and expert systems to control the environmental impacts of economic growth. As Ulrick Beck (1998, p. 26) suggests, ecological modernization is unlikely to provide a solution beyond "cosmetic ecology on grand scale". Despite the extensive criticism of the "three pillar" or "magic triangle" (Campbell, 1999) approach to sustainable development, it continues to frame a large part of the environmental discourses in planning. Moreover, the emphasis has gradually moved away from balancing the three pillars towards seeking sustainable "economic" growth (Davoudi *et al.*, 1996). Blühdorn & Welsh (2007, p. 187) call this trend "the hegemony of eco-economic 'win-win' thinking". A kind of thinking that can also be traced in the climate mitigation discourses of, for example, two national planning statements where promotion of renewable energy (ODPM, 2004a) or biodiversity (ODPM, 2005a) are framed and justified in terms of their potential not only for reducing greenhouse gas emissions (GHG) but for diversification of rural "economies".

The more radical interpretations of sustainability, notably the reflexive ecological modernization (Hajer, 1995), are difficult to find in the environmental narratives of planning. This seems to reflect the wider "politics of unsustainability" which, according to Blühdorn and Welsh (2007, p. 198), is unfolding as a result of two simultaneous trends: firstly, "a general acceptance that achievement of sustainability requires radical change in the most basic principles of late-modern societies" and secondly, a general "consensus about the non-negotiability of democratic consumer capitalism". Both trends are happening "irrespective of mounting evidence of unsustainability" (Blühdorn and Welsh, 2007) and a growing awareness about climate change. The latter has demonstrated, perhaps more than any other environmental concerns, our intimate and inescapable interdependencies with nature. It is a powerful reminder of the complexities of the human-nature inter-relationship and the fallacy of the modernist assumption about the ability of human to tame and order nature for their own exploitation with little or no consequences.

2.8 *Environment as Risk*

Climate change has introduced two new concepts, mitigation[6] and adaptation,[7] each with distinctly different implications for environmental discourses in planning. As will be discussed below, the discourse of climate "mitigation" has reinforced the environmental aspects of sustainable development through its emphasis on efficient use of energy and reduction in non-renewable energy sources. The discourse of climate "adaptation", on the other hand, seems to have marked a departure from sustainable development in the sense that it considers the environment as a "risk to", rather than as "an asset for", human well-being. Seeing the environment as a natural hazard and a threat against which resilience should be built is an entirely new perspective in planning (see Davoudi, 2012 for a critical review of the concept of resilience).

2.8.1 *The mitigation discourse.* Although the term climate mitigation did not appear in planning until the early 2000s, throughout the 1990s reduction of GHG emissions through, for example, reducing the need to travel was a frequently rehearsed planning discourse (Owens & Cowell, 2002). By the mid-2000s, changes to the planning system[8] and the development of climate change policy at the international and national levels[9] placed climate mitigation firmly at the centre of the planning agenda. As part of the statutory purpose of planning to contribute to sustainable development, planners have to "ensure that development plans contribute to global sustainability by addressing the causes and potential impacts of climate change" (ODPM, 2005b, p. 13). By 2006, all English regions had a climate change regional strategy in place which provided the basis for climate change policies in local planning documents. As a result, mitigation narratives have been increasingly interwoven into, and sometimes eclipsing, the discourses of sustainability to the extent that climate change has now become emblematic in environmental discourses at all levels of planning. At the European level, for example, Sykes and Fischer (2009) mention the escalating appearance of the term "climate change" in the European spatial planning documents from nine in the ESDP in 1999 (CEC, 1999) to 26 in its replacement in 2007. As Hajer (1995, p. 5) suggests, climate change has become "a 'metaphor' for the environmental problematique at large". However, despite being the political flavour of our time two issues are still lurking in the background and demand attention. The first is the continuation of the economic imperatives and the appropriation of climate change concerns towards an emphasis on economic growth through, for example, the "more secure supplies of energy … and harnessing of new forms of technological innovation … " (Jordan, 2009, p. 764). The second is the intensification of "all the unresolved tensions that the concept of sustainable development was originally designed to confront" (Meadowcroft, 2007).

2.8.2 *The adaptation discourse.* Adaptation to climate change did not initially attract as much international attention as climate mitigation for, arguably, two reasons: firstly, the fear that it might distract governments' attention from carbon reduction, and secondly, the belief that adaptation was a developing countries' problem (Swart & Raes, 2007). Despite its late arrival, climate adaptation is catching up and infusing the environmental discourses of planning with a multitude of new vocabularies. A key feature of the adaptation narrative is its construal of the environment as natural hazard and the framing of our relation to nature in terms of risk and security. So, in contrast with the discourse of sustainability it portrays the environment not so much in terms of assets to be "sustained for" human benefits, but in terms of threats "against which" human well-being should be safeguarded. As such it can be seen as a hark back to a pre-modern conception of human-nature relationship which was centred on what nature does to us, rather than what we do to nature. The former view dominated the early hazard research and disaster risk management which considered nature's forces, rather than human action, as being responsible for putting people at risk. While this perspective has been challenged (Blaikie *et al.*, 1994) it continues to influence the framing of climate adaptation (Fünfgeld & McEvoy, 2011).

The departure of the adaptation discourse from sustainable development agenda has a number of other dimensions. Firstly, contrary to the rhetoric of sustainable development which imbued environmental discourses of plans with optimism of win-win solutions, adaptation discerns a subtle sense of pessimism (Blowers *et al.*, 2009). The emerging negative and defensive language is reflected, for example, in a shift in focus in the Environment Agency's latest strategic plan. While the discourses of its previous plans centred on positive measures for "a better quality of life for people and an enhanced environment for wildlife", its latest plan pledges to "clean up rivers, defend more properties against flooding, reduce emissions and tackle the causes and consequences of climate change" (cited in Vidal *et al.*, 2010). Secondly, although the rhetoric of sustainable development projects an image of a secure and sustainable future, the language of adaptation conjures up images of a future characterized by the inevitability of climate risks and insecurities. Within this narrative, unpredictability and uncertainty replace the (false) sense of certainty and the overrated human ingenuity and technology. This, however, is not to suggest that the adaptation discourse brings a reversal of the modernist "quest to control" which itself stemmed from real and perceived fears and anxieties about complexity and uncertainty (Schumaker, 1978). On the contrary, climate uncertainties have heightened the quest for control which is reflected, for example, in the numerous computer models that are being generated to bring uncertainty into the range of the so called "predictable probability" (Johnston *et al.*, 2000). An example of this mixture of pessimism about the future and optimism about our ability to control it can be found in the London Climate Change Adaptation Plan. It first devotes a long section under the heading: "Catastrophic climate change" (GLA, 2010, p. 21) and reveals that, "London is ranked the ninth most vulnerable megacity on the risk register of natural hazards for the world's 50 megacities" (GLA, 2010, p. 109). Ironically, this rather unsettling piece of news is presented under the sub-heading of: "A safe place for business"! It then reassures people, or more precisely the businesses, that "London is less vulnerable than some of its principal world city competitors (New York and Tokyo)" (GLA, 2010, p. 109), as if risk becomes acceptable as long as others are more at risk than London. The reassurances are reinforced by Mayor's foreword stating that, "we can approach this task (which he calls "to weather proof our city") with optimism rather than gloomy defeatism" (GLA, 2010, p. 6). Swyngedouw (2007, p. 19) refers to this paradox as "an unbridgeable gap" between our "dominant view of Nature" as a predictable process moving towards equilibrium which has been disturbed by human action but can also be rectified by human action, and "the acting-out of natures" as an unpredictable and chaotic set of processes.

The third dimension of the departure from sustainable development is the emphasis on risk-based approaches to planning. While disaster risk management is not new in government policy and is certainly not exclusive to climate adaptation (Coaffee *et al.*, 2008; Pidgeon & Butler, 2009) its articulation in planning is relatively new. Its first noticeable appearance was in 2004 in a government's best practice advice (ODPM, 2004b) where an entire section of the document was devoted to the topic. Since then the language of risk and insecurity is increasingly interlaced with other environmental discourses in planning. Seeing the environment as dangerous and hazardous is most prominent in planning documents that have to deal with flooding, as might be anticipated. Since 2000, three versions of the national policy guideline on flood-risk management have been produced. The latest states that, "A risk-based approach should be adopted at all levels of planning" based on a "sources-pathway-receptors" approach. The "receptors" include, presumably in the order of significance, "people, property, infrastructure, habitats and statutory sites" (DCLG, 2006, revised 2010). The approach implies that assessing the impact of development "on" the environment is now accompanied with assessing the risk "of" the environment for development.

The emerging risk-laden meaning of the environment in planning is in line with the wider and dominating global discourse of "securitization" (Aradau, 2009) which covers an expanding range of concerns such as, climate, energy, food, asset and life. "A world increasingly subject to risks" is seen by some as the latest changes that have occurred to the concept of sustainable development (Hamdouch & Zuindeau, 2010, p. 430). Blowers *et al.* (2009, p. 313) suggest that it personifies a "shift from an environmental politics of cooperation and consensus to a politics of securitisation", with the emphasis moving from local to national level of decisions-making. Others concur, and argue that the over-emphasis on security may lead to diminution of democracy (Coaffee *et al.*, 2008) and legitimization of "the suspension of democratic safeguards and the uncoupling of checks and balances" (Blühdorn & Welsh, 2007, p. 191) on the ground that urgent action is required. A subtle reflection of this can be found in some climate adaptation strategies which invite public comments yet without making an explicit commitment to take them on board. In the London adaptation plan, for example, the Mayor calls "on people to tell" him "what their solutions are", yet makes it clear that only "the best of these will inform the final version of the Strategy" (GLA, 2010, p. 7). Several months after the consultation period, there was still no sign of the analysis of the received comments on the GLA website.

Furthermore, the adaptation "turn" in planning with its emphasis on risk is renewing a number of outmoded approaches to planning. One example is the return to the 1960s' technical rational decision-making. This is reflected in the 2004 best practice advice (mentioned above) which advocates seven logical steps for making "rational" adaptation decisions.[10] Another related trend is the emphasis on the use of engineering-based modelling in adaptation decisions; a trend which can challenge the qualitative and judgemental approaches to planning decisions in favour of quantitative and calculative ones. Increasingly, the risk-based approaches are articulated not only as technical knowledge practices, but also as a particular form of decision-making and governance. Hence, risk tools are being adopted not just as technical aids to decision-making, but also as political strategies for climate change mitigation and adaptation (Pidgeon & Butler, 2009).

3. Conclusion

"There is no singular 'nature' as such, only a diversity of contested natures; and each such nature is constituted through a variety of socio-cultural processes from which such nature cannot be plausibly separated" (MacNaghten & Urry, 1998, p. 1).

Environment defies simple definition. Its meanings are multiple, conflicting and contested. In the last decades planning institutions and practices have provided key sites

Table 2. Meanings of "the environment" in English planning

Meaning of the environment in planning	View of human-nature relationship	Dominant approach	Rooted in	Discursive focus	Manifestation in planning (example of planning tools used)
Local amenity	Anthropocentric. Nature as a refuge for human enjoyment	Welfare-oriented aesthetic utilitarian	Green romanticism (nineteenth century)	Protection of intangible local amenities	Utilization of natural beauty as a backcloth for development Separation of urban from rural (green belt)
Heritage landscape	Anthropocentric. Human as the steward of a "green" nature	Aesthetic utilitarian	Elitist preservation of a pre-industrial past (nineteenth century)	Preservation of the "best" of heritage landscape	Hierarchical categorization of bounded parcels of landscape (site designation)
Nature reserve	Bio-centric. Human as part of nature and the web of life	Ecological	Radical environmentalism (1960s/1970s)	Enhancement of nature for nature's sake	Making connections (green corridors) Alongside hierarchical ordering (sieve mapping)
Storehouse of resources and functions	Anthropocentric. Human as the master of a "clockwork" nature	Functional utilitarian	Scientific revolution and mechanistic view of the world (seventeenth/eighteenth century)	Exploitation of natural resources for economic growth	Using nature as a container of materials, scientific repository and a sink for absorbing pollution/waste Separating polluters from people (zoning)
Tradable commodity	Anthropocentric. Nature as a "stock of capital" for human consumption	Market-oriented utilitarian	Neo-liberalism (1980s)	Bargaining with nature	Trading nature for development gain (planning gain)
Problem	Anthropocentric. Human action as a remedy for natures' problems	Administrative rational: "leave it to expert"	Environmental awareness (1960s/1970s)	Problem-solving Rectifying environmental problems without socio-economic restructuring	Assessing and reducing environmental impacts of development (EIA)
Sustainability	Anthropocentric. Nature as an asset to be safeguarded for future generations	Ecological modernization	Forestry practices (eighteenth century) Brundtland report (1987)	Achieving win-win solutions	Balancing social, economic and environmental goals (SEA)
Risk	Anthropocentric. Nature as a threat against which future generation should be safeguarded	Technical-rational risk assessment	Climate change adaptation (2000s)	Securitization, building resilience	Assessing and reducing environmental risks to development (climate proofing)

for the development and contestation of the changing environmental discourses. The narratives of the development plans in particular have reflected and reinforced the evolving meanings of the environment. Based on the analysis of a sample of plans produced in 2000s and while recognizing the nuances of the environmental discourses, this paper has identified eight broad meanings of the environment in planning. Over time new discourses have been added to planners' repertoire to create what is now a multi-layered narrative of the environment in contemporary plans. The environment is seen simultaneously: as local amenity, as heritage landscape, as nature reserve, as storehouse of resources, as tradable commodity, as problem, as sustainability and as risk (see Table 2 for a summary). Some of these discourses may complement each other (such as local amenity and functional services), others are often in conflict (such as nature reserve and storehouse of material resources) and most are in competition for setting the agenda for policy and action. Their formation, continuation and transformation are contested political acts, infused with power relations.

While each narrative has its own particular history and trajectory, what binds them together is a deep-seated anthropocentric view of nature, and of human as its steward or even its master. The most explicit manifestation of this human-centred view is the persistent utilitarian treatment of the environment as a storehouse of resources and functions for human exploitation. Interlaced with this functional view is the conservation narrative which has centred on protecting the "best" of nature for human enjoyment. The "best" is constructed through a system of hierarchical designations which conceives the environment as a series of discrete parcels, and values each parcel according to its position on the hierarchy. Dividing nature into fixed entities (such as air, water, land) or bounded spaces (such as SSSIs, AONBs) constitutes a key planning "technology" for managing and controlling nature; a practice which is also embedded in planning's rational administrative responses to solving environmental "problems". Although these, coupled with an intensification of EU-driven environmental regulations and impact assessments, have expanded the discursive space for the environmental narratives in planning, it has not led to a shift towards more radical bio-centric views of the environment. Even the sustainability agenda has been subsumed into a weak ecological modernization whose positive-sum narrative of balancing competing economic, social and environmental goals has become a powerful rhetoric in planning. However, this positive image of a "sustainable" future which imbued planning for nearly two decades is now being challenged by the growing awareness of the inevitability of climate change. It appears that "the term environment is being subsumed under the hegemonic imperative of climate change ... " (Whatmore, 2008, p. 1777). As a result, environment has come to be seen as natural hazard and a risk to human well-being. It is no longer seen only as an asset to be sustained for future generations, but also as something that presents a threat to them. Environment is now increasingly framed in the language of security, risk and resilience (see Davoudi, 2012 for a critique of resilience). Risk management and "emergency" planning sit alongside "spatial" planning in this climate of heightened securitization. However, the adaptation "turn" in planning and its associated discourses of risk and security may be short lived, given the current relaxation of the centrally enforced requirements for producing adaptation plans. In the future, what new meanings of the environment emerge in planning discourses depend fundamentally on what kind of natures are imagined and what type of human-nature relationships are pursued.

Acknowledgements

This paper is partly based on the research undertaken for the Arcadia Project (Adaptation and Resilience in Cities: Analysis and Decisions-making using Integrated Assessment, award number EP/G060983/1). I would like to thank the Engineering and Physical Science Research Council for their financial support. I would also like to thank Dr Elizabeth Brooks at Newcastle University for her invaluable assistance in the review of the literature and the analysis of the planning documents, and the two reviewers for their constructive comments.

Notes

1. The focus of the paper is on the planning system in England which following the devolution of planning powers differs from the rest of the UK.
2. Discourse in this paper is defined as a shared narrative which enables those who subscribe to them make sense of fragmented information (Dryzek, 1997).
3. It is now called the Campaign to Protect Rural England.
4. Along with the work of Galileo, Kepler and Copernicus on mechanics and cosmology.
5. "Sustainable development is development that meets the needs of this generation without compromising the ability of future generations to meet their needs" (WCED, 1987, p. 8).
6. Defined as the need to drastically reduce the anthropogenic (human-induced) greenhouse gas emissions.
7. Defined as the need to adapt to the impact of climate change.
8. Through the Planning and Compulsory Purchase Act 2004.
9. Such as the UK Climate Impact Programme (2006), the EU White Paper on Adapting to Climate Change (2009) and the UK Climate Change Act (2008).
10. These include "Identify the scope of the policy; Establish criteria for policy making and exposure unit, Assess risk, Identify options, Appraise options, Formulate policy, and Implement, monitor and review" (ODPM, 2004b).

References

Abercrombie, P. (1933) *Town and Country Planning* (Oxford: Oxford University Press).

Aradau, C. (2009) Climate emergency: Is securitisation the way forward? in: D. Humphreys & A. Blowers (Eds) *A Warming World, Vol 1 of Course DU311, Earth in Crisis*, pp. 165–205 (Milton Keynes: The Open University).

Barnes, T. & Duncan, J. (1992) *Writing Worlds: Discourse, Text and Metaphor in the Representation of Landscape* (London: Routledge).

Beck, U. (1998) *Democracy Without Enemies* (Cambridge: Polity Press).

Blaikie, P., Cannon, T., Davis, I. & Wisner, B. (1994) *At Risk: Natural Hazards, People's Vulnerability, and Disasters* (London: Routledge).

Blowers, A., Boersema, J. & Martin, A. (2009) Whatever happened to environmental politics? *Town and Country Planning*, 78(7/8), pp. 311–315.

Blühdorn, I. & Welsh, I. (2007) Eco-politics beyond the paradigm of sustainability: A conceptual framework and research agenda, *Environmental Politics*, 16(2), pp. 185–205.

Botkin, D. B. (1990) *Discordant Harmonies, A New Ecology for the Twenty-First Century* (New York: Oxford University Press).

Bowler, P. (1992) *The Fontana History of the Environmental Sciences* (London: Fontana Press).

Campbell, S. (1999) Planning, green cities, growing cities, just cities? Urban planning and the contradictions of sustainable development, in: D. Satterthwaite (Ed.) *Sustainable Cities*, pp. 251–274 (London: Earthscan Publications).

CBC (Chelmsford Borough Council) (2008) *Core Strategy and Development Control Policies, Chelmsford Borough Council Local Development Framework 2001–2021* (Chelmsford: CBC).

CEC (Commission of the European Communities) (1999) *European Spatial Development Perspective* (Luxemburg: CEC).

Coaffee, J., Murakami Wood, D. & Rogers, P. (2008) *The Everyday Resilience of the City: How Cities Respond to Terrorism and Disaster* (Basingstoke: Palgrave Macmillan).

Davoudi, S. (1997) Environmental considerations in minerals planning, theory versus practice, in: D. Borrie, A. Khakee & C. Lacirignola (Eds) *Evaluating Theory-Practice and Urban-Rural Interplay in Planning*, pp. 157–165 (The Netherlands: Kluwer Academic Publisher).

Davoudi, S. (2000a) Planning for waste management: The changing discourses and institutional relationships, *Progress in Planning*, 53(3), pp. 165–216.

Davoudi, S. (2000b) Sustainability: A new vision for the British planning system, *Planning Perspectives*, 15(2), pp. 123–137.

Davoudi, S. (2012) Resilience, A bridging concept or a dead end? *Planning Theory and Practice*, forthcoming.

Davoudi, S. & Layard, A. (2001) Sustainable development and planning: An introduction to concepts and contradictions, in: A. Layard, S. Davoudi & S. Batty (Eds) *Planning for a Sustainable Future*, pp. 7–19 (London, Spon).

Davoudi, S., Hull, A. & Healey, P. (1996) Environmental concerns and economic imperatives in strategic plan-making, *Town Planning Review*, 64(4), pp. 421–436.

DCLG (Department for Communities and Local Government) (2006, revised 2010) *Planning Policy Statement 25: Development and Flood Risk* (London: TSO).

DCLG (2007) *Planning Policy Statement 22: Renewable Energy* (London: DCLG).

DoE (Department of the Environment) (1985) *Circular 14/85 Development and Environment* (London: HMSO).

Dryzek, J. (1997) *The Politics of the Earth* (Oxford: Oxford University Press).

Fünfgeld, H. & D. McEvoy (2011) Framing Climate Change Adaptation in Policy and Practice. Working Paper 1. Melbourne: Victorian Centre for Climate Change Adaptation Research. http://www.vcccar.org.au/files/vcccar/Framing_project_workingpaper1_240611_0.pdf (accessed 5 December 2011).

Gibbs, D., While, A. & Jonas, A. (2007) Governing nature conservation: The EU habitats directive and conflicts around estuary management, *Environment and Planning A*, 39(2), pp. 339–358.

GLA (Greater London Authority) (2009) *The London Plan: Spatial Development Strategy for Greater London - Consultation Draft Replacement Plan* (London: GLA).

GLA (2010) *The Draft Climate Change Adaptation Strategy for London. Public Consultation Draft* (London: GLA).

GOEE (Government Office for the East of England) (2008) *The East of England Plan: The Revision to the Regional Spatial Strategy for the East of England* (London: DCLG).

GONW (Government Office for the North West) (2008) *North West of England Plan: Regional Spatial Strategy to 2021* (London: TSO).

Hajer, M. A. (1995) *The Politics of Environmental Discourse: Ecological Modernisation and the Policy Process* (Oxford: Clarendon Press).

Hamdouch, A. & Zuindeau, B. (2010) Sustainable development 20 years on: Methodological issues, practices and open issues, *Journal of Environmental Planning and Management*, 52(4), pp. 427–438.

Healey, P. & Shaw, T. (1994) Changing meanings of "environment" in the British planning system, *Transaction of the Institute of British Geographer*, 19(4), pp. 425–438.

Hebbert, M. (2009) The three P's of place-making for climate change, *Town Planning Review*, 80(4–5), pp. 35–370.

Hollis, M. (2003) *The Philosophy of Social Sciences* (Cambridge: Cambridge University Press).

Johnston, R. J., Gregory, D., Pratt, G. & Watts, M. (Eds) (2000) *The Dictionary of Human Geography,* 4th ed. (Oxford: Blackwell).

Jordan, A. (2009) Revisiting . . . the governance of sustainable development: Taking stock and looking forwards, *Environment and Planning C*, 27(5), pp. 762–765.

Lafferty, W. & Meadowcroft, J. (Eds) (1996) *Democracy and the Environment* (Cheltenham: Edward Elgar).

LCC (Lancaster City Council) (2008) *Local Development Framework Core Strategy 2003–2021 (Adopted July 2008)* (Lancaster: LCC).

LBM (London Borough of Merton) (2010) *Local Development Framework Core Strategy (Pre-submission Representation Publication)* (London: LBM).

LCC (Lancashire County Council) (1974) *The Outline Plan for the Central Lancashire New Town* (Preston: LCC).

LCC (Lancashire County Council) (1994) *Lancashire Structural Plan: Greening the Red Rose County* (Preston: LCC).

Lowe, P. and Goyder, J. (1983) *Environmental Groups in Politics* (London: Allen and Unwin).

Macnaghten, P. & Urry, J. (1998) *Contested Nature* (London: Sage).

Marsden, T., Murdoch, J., Lowe, P., Munton, R. & Flynn, A. (1993) *Constructing the Countryside* (London: UCL Press).

Marshall, P. (1994) *Nature's Web, Rethinking our Place on Earth* (New York: Paragon House).

Meadowcroft, J. (2007) National sustainable development strategies: Features, challenges and reflexivity, *European Environment*, 17(3), pp. 152–163.

Meadows, D., Randers, J. & Behrens, W. (1972) *The Limits to Growth, A Report for the Club of Rome's Project on the Predicament of Mankind* (London: Pan).

Newby, H. (1990) Ecology, amenity and society: Social science and environmental change, *Town Planning Review*, 61(1), pp. 3–13.

ODPM (Office of Deputy Prime Minister) (2004a) *Planning Policy Statement 22: Renewable Energy* (London: HMSO).

ODPM (2004b) *The Planning Response to Climate Change: Advice on Better Practice* (London: ODPM).

ODPM (2005a) *Planning Policy Statement 9 Biodiversity and Geological Conservation* (London: HMSO).

ODPM (2005b) *Planning Policy Statement 1: Delivering Sustainable Development* (London: HMSO).

Owens, S. & Cowell, R. (2002) *Land and Limits: Interpreting Sustainability in the Planning Process* (London: Routledge).

Pearce, D. & Barbier, E. (2000) *Blueprint for Sustainable Development* (London: Earthscan).

Pearce, D., Mmarkandya, A. & Barbier, E. (1989) *Blueprint for a Green Economy: Report for the Department of the Environment* (London: Earthscan).

Pidgeon, N. & Butler, C. (2009) Risk analysis and climate change, *Environmental Politics*, 18(5), pp. 670–688.

Schumaker, P. D. (1978) The quest for control: A critique of the rational-central-rule approach in public affairs, *The Journal of Politics*, 40(1), pp. 239–242.

Sharp, T. (1932) *Town and Countryside* (Oxford: Oxford University Press).

Swart, R. & Raes, F. (2007) Making integration of adaptation and mitigation work: Mainstreaming into sustainable development policies? *Climate Policy*, 7(3), pp. 288–303.

Swyngedouw, E. (2007) Impossible "sustainability" and the postpolitical condition, in: R. J. Kruger & D. Gibbs (Eds) *The Sustainable Development Paradox, Urban Political Economy in the United states and Europe*, pp. 13–41 (New York: Guilford Press).

Sykes, O. & Fischer, T. (2009) The territorial agenda of the European Union: Progress for climate change mitigation and adaptation? in: S. Davoudi, J. Crawford & A. Mehmood (Eds) *Planning for Climate Change: Strategies for Mitigation and Adaptation for Spatial Planners*, pp. 111–125 (London: Earthscan).

Verhagen, F. C. (2008) Worldviews and metaphors in the human-nature relationship, an ecolinguistic exploration through the ages, *Language and Ecology*, 2(3), Available at http://www.ecoling.net/worldviews_and_metaphors_-_final.pdf (accessed 6 December 2011).

Vidal, J., Carrell, S. & Jowitt, T. (2010) Crown jewels of Britain's landscape could be sold off. *The Guardian*, August 14. Available at http://www.guardian.co.uk/environment/2010/aug/13/crown-jewels-britains-landscape-sold (accessed 7 September 2010).

WCED (World Commission on Environment and Development) (1987) *Our Common Future* (Oxford: WCED).

Whatmore, S. (2008) Editorial: Remaking environment: History, practices, *Policies, Environment and Planning A*, 40(8), pp. 1777–1778.

Whatmore, S. & Boucher, S. (1993) Bargaining with nature: The discourse and practice of "environmental planning gain", *Transactions of the Institute of British Geographers*, 18(2), pp. 166–178.

Williams, R. (1972) Ideas of nature, in: J. Benthall (Ed.) *Ecology, the Shaping Enquiry*, pp. 147–161 (London: Longman).

Worster, D. (1994) *Nature's economy, a history of ecological ideas*, 2nd ed. (New York: Cambridge University Press).

Rotterdam: A City and a Mainport on the Edge of a Delta

HAN MEYER*, ANNE LOES NILLESEN* & WIL ZONNEVELD**

*Faculty of Architecture, Delft University of Technology, Julianalaan 134, 2628 BL Delft, The Netherlands,
**OTB Research Institute for the Built Environment, Delft University of Technology, Jaffalaan 9, 2628 BX Delft, The Netherlands

ABSTRACT *Within Europe, Rotterdam is by far the largest port and supplier of fossil energy sources. City and port have a "sandwich" position in the low lands between a sea with a rising level and rivers with increasing peak discharges. It is certainly no exaggeration to say that sustainability forms a matter of life or death for Rotterdam as a Delta City. The question of a sustainable Rotterdam or not is related to the following issues: (1) water management (preventing hazards; the restoration of the estuary; salinization); (2) urban renewal; (3) the spatial and climate footprint of the ever-growing port and (4) energy transition. Currently all these issues are dealt with largely independently of one another. For a genuinely sustainable future, linkages have to be made between strategies, projects and actors.*

1. Introduction

The city of Rotterdam can ascribe its economic "raison d'être" and power as well as its increasing vulnerability to the same source, its position in the Rhine–Meuse delta, at the edge of the north-western European "megalopolis". Being the largest European port and supplier of fossil energy sources, and struggling with its "sandwich" position in the low lands between a sea with a rising level and rivers with increasing peak discharges, "sustainability" is a far from meaningless term in this city. We distinguish four main topics which can be regarded as questions of life or death for this city.

The first topic concerns the way the city and the port can survive during changing conditions of sea, rivers and rainfall. With a dense concentration of people, capital and economic activities (the Rotterdam region is responsible for 8% of the GDP of the Netherlands), the first question is how the risk of flooding can be reduced to a minimum. The central water problem in this delta region is the danger of flooding, but related to this, questions about salinization, soil subsidence and rainwater management need to be taken into account.

The second topic concerns the strategic position of the city as a European mainport in relation to energy transition. If the city aims to continue exploiting this strategic position, the port policy should consider how the port can anticipate the future transition of energy sources. Future scenarios concerning this energy transition should result in new visions and strategies concerning land use, accessibility, networks, supporting economies, etc. Moreover, being one of the most important contributors to CO_2 emissions, the port is being forced to look for possibilities to reduce these emissions substantially.

Third, the development of a new flood-risk strategy, combined with a substantial transformation of the port, will result in new possibilities for the urbanization of docklands and riverfronts. Altogether we are talking about huge areas of hundreds of hectares which will be available for urbanization in the future. This provides an opportunity for the city to develop an urban design policy, emphasizing new spatial and functional qualities which might be important to stimulate specific social and economic processes.

Fourth, until recently, neither flood-risk management nor the port economy was considered as very good for nature and ecology. Land reclamations, the construction of many dikes and dams, and the expansion of the port industry resulted in the disappearance of wetlands and estuaries, a radical change in the ecosystem of the delta and a dramatic loss of bio-diversity. A new awareness of the importance of ecological sustainability, embodied in European rules and civic pressure groups, has forced the city and the port authorities to develop new strategies concerning flood-risk management and port development, contributing to the repair of the deltaic natural system.

Certain issues which are important for Rotterdam and Rijnmond and could be considered as part of a wider interpretation of sustainability are not considered here. For example, the social–economic problems in the greater Rotterdam area are greater than anywhere else in the Netherlands. The economic (industrial) history has left a legacy of problems: a relatively one-sided economy, high concentrations of social and economic deprivation and a massive urban renewal problem. Since 2006, the Rotterdam South area has been one of the main targets of a national government programme to improve social and economic conditions in problematic urban areas (Van den Brink, 2007).

The city administration of Rotterdam is quite aware of the importance of the development of an effective policy concerning the mentioned aspects of sustainability. Because of this reason, the Rotterdam Climate Initiative (RCI) programme was started in 2008. The aim is for the city to become a "worldwide benchmark" for dealing with climate change and to make the city "climate-proof" by 2025.[1] Water and water management as well as energy transition and the reduction of CO_2 emissions play an important role in this programme. An important question that remains is whether we can really expect an innovative and comprehensive approach by the city.

This paper starts with a general introduction to the origins of the Delta City known as Rotterdam and Rijnmond—we intermittently use these two names, in most cases referring to the wider Rotterdam area: the city and the dozen or so municipalities on either side of the Nieuwe Waterweg. Section 3 looks at how the Rijnmond mainport system has evolved and the often non-critical—from a present day perspective—political decisions which were made in order to achieve a continuous and almost endless expansion of the Rotterdam harbour. We focus on the prime sustainability and climate change issues facing the Rijnmond Delta City independent from the water system. In Section 4, we look at the latest expansion of the Rotterdam port and how the Maasvlakte 2 project—the name given to the new port area—has become immersed in discourses about sustainability, delta nature and climate change. In Section 5, we change our perspective to the water-related problems that the Rotterdam Delta faces. Section 6 focuses on the possibilities of new types of urban environments because of the new strategies concerning flood management and port policy. In Section 7, we focus on the possibility of developing comprehensive strategies, based upon scenarios concerning flood management and port development. We round off with conclusions and discuss some of the future key sustainability and climate change issues related to delta and (main)port development.

2. One Delta City with Two Faces

The spatial development of the city of Rotterdam has been determined strongly by its position at the edge of the southwest delta area of the Netherlands: this position defines its economic position and spatial structure and has influenced the social diversification of the city. This "edge position" means that Rotterdam is settled on two different ground structures, separated by the Nieuwe Maas river. North of this river, the ground structure is mainly peat, part of the "peat continent" of central Holland (the area of the two western provinces with "Holland" their names), which stretches from the river Maas to the banks of the IJ north of the historic core of Amsterdam. This peat area was the first area to be surrounded by dikes in the thirteenth century, and since then has been relatively safe for human settlement (Van de Ven, 2004). This enclosure still plays an important role in the water management and flood control of Holland as "dike-ring 14", as part of the system of dike rings which is the fundament of water management and flood control in the Netherlands (Figure 1). Dike-ring 14 surrounds the area with the highest safety standard and created the condition for a process of intensive urbanization (De Vries & Van de Woude, 1997), resulting in what we call the "Randstad" today.

South of the river Nieuwe Maas we find the actual delta, composed of a number of estuaries, islands and peninsulas, with a ground structure which is mainly clay. Until the nineteenth century, the city of Rotterdam was built on the north bank of the river, in the peat area and mainly safely behind the dike. Part of the city, especially the port area, was built on raised areas outside the dikes. From the mid-nineteenth century, two important spatial developments took place.

First, the accessibility of the Port of Rotterdam was improved by digging the "Nieuwe Waterweg" which started in 1863. The Port of Rotterdam struggled with an increasing inaccessibility because of sedimentation processes in the main channels of the delta. This sedimentation was caused by the rivers as well as by the tidal currents of the sea. By digging a new channel from Rotterdam to the sea, parallel to the southern dike of Dike-ring 14, a new main discharge channel for the river was created. This channel resulted in a higher speed of the river currents and consequently in an increased discharge of sediments and cleaning of the river bed. The new channel combined two goals: it contributed to the control of the water levels of the principal rivers in the Netherlands, and it created a new deep-water access to the Port of Rotterdam. But it also resulted in the influence of the sea increasing in the Rotterdam region itself, in terms of bigger tidal differences of the river water level and an increased vulnerability to storm surges, as well as in terms of increased salinization of the whole region (Van de Ven, 2004).

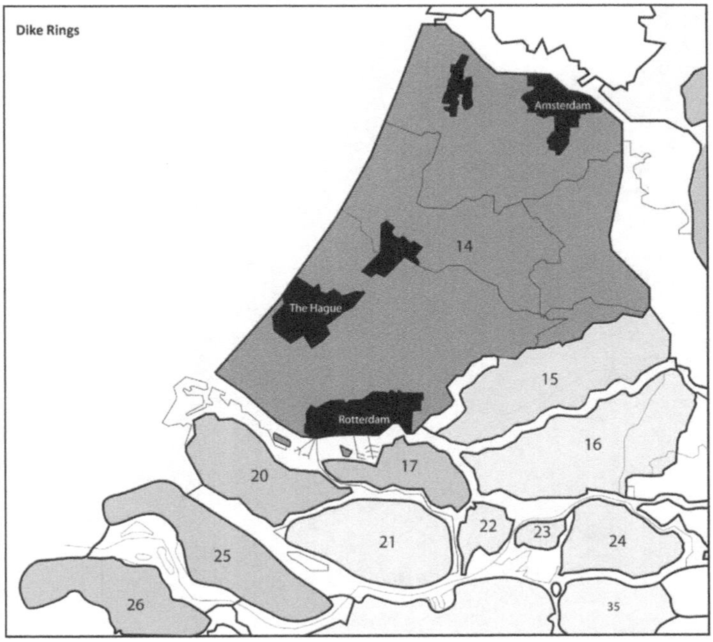

Figure 1. Dike rings in the western part of the Netherlands
Source: Authors.

The second important spatial development concerns the extension of the city and port on the left banks of the river. During the nineteenth and twentieth centuries, the two parts of the city on both sides of the river banks developed in two different ways. The north bank, with the historic city, maintained its role as the city centre and developed some prosperous urban districts. The south bank is still part of the deltaic landscape and shows a more fragmented character; the former structure of the area as a conglomeration of small islands is still recognizable in the urban pattern of this part of the city (Palmboom, 1987) (Figures 2 and 3). The first generations of residents of this part of the city worked mostly in port-related industries (Bouman & Bouman, 1952). For a large part, this was related to the fact that the left, southern riverbank became the main territory of port development of the nineteenth and twentieth centuries.

At the moment, the municipal territory of Rotterdam covers 314 km^2, with ca. 600,000 inhabitants. The metropolitan region has ca. 1.2 million inhabitants. The city territory includes a port area of 105 km^2 (Figure 4), which is in the unembanked areas. Several urban districts, most of them built on former port areas, can also be found in unembanked areas.

3. Conflicts about the Growing Territorial Footprint of Rotterdam Mainport

The open connection between the sea, the port and its hinterland has been a crucial factor in the development of the Port of Rotterdam. It is generally accepted that the Rotterdam port owes its present position to its favourable location in northwest Europe in general and the excellent water connections to Germany in particular.

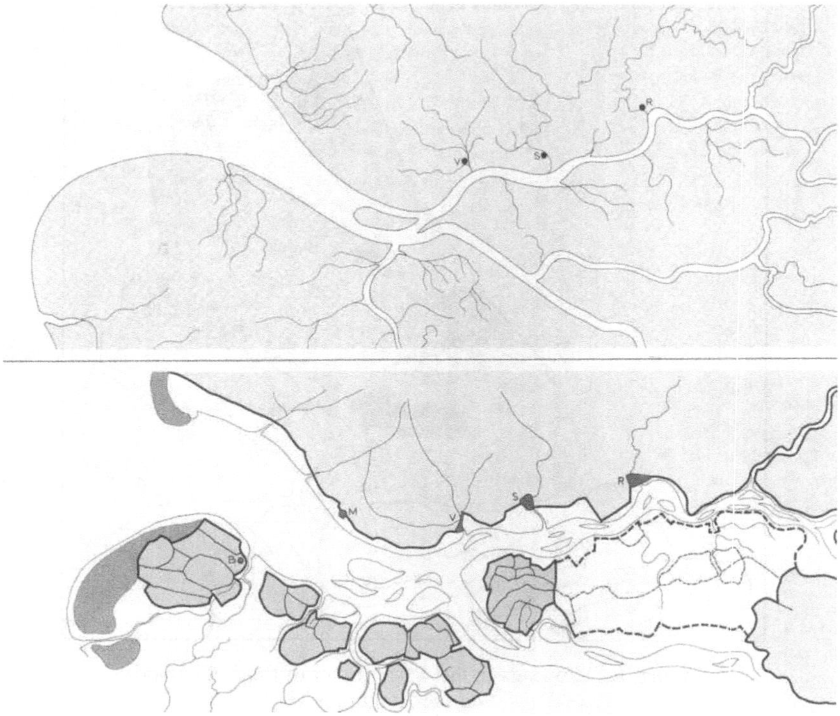

Figure 2. The Rotterdam region before 1100, dominated by rivers and creeks (top); and around 1400, with a dike along the north bank of the river and islands with dike rings in the south (bottom)
Source: Palmboom (1987)

3.1. *A Transportation Hub*

Although no longer the largest seaport in the world—this position has been taken by Asian seaports—figures on the volume of flows at the Rotterdam mainport are still extremely impressive. The latest available figures for 2009 show that Rotterdam in terms of transhipment is more than twice the size of the second port of Europe, Antwerp: 387 million metric tonnes compared with 157.6 for Antwerp. Most of this volume, about 71%, goes into the harbour; 29% is transferred to other sea vessels (transhipment). When it comes to the number of containers the differences with its main competitors are less. Rotterdam handles 9,743,000 TEUs (Twenty-Feet Equivalent Units) while the figure for Antwerp—the second European harbour in this respect as well—is 7,310,000 TEUs. Hamburg is close on the heels of Antwerp with 7,008,000 TEUs (Port of Rotterdam 2009).

Although the market position of Rotterdam in the so-called Le Havre-Hamburg range—the range of harbours catering for the same hinterland—is gradually diminishing, the volume of goods and consequently the number of transportation units are enormous. In 2009, no less than 29,000 sea vessels entered the Rotterdam harbour areas to deliver or take up goods and containers.

From the perspective of the ecological footprint of the harbour, it is interesting to look at other figures: what does the mainport position of the Rotterdam harbour area mean in terms of flows running through the metropolitan region: trains, lorries and inland

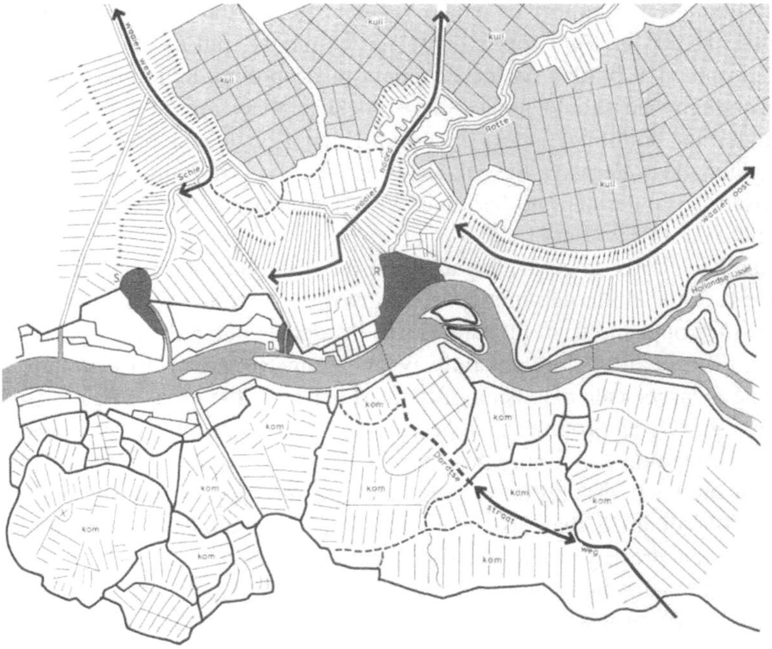

Figure 3. The spatial structure of Rotterdam: north of the river defined by long lines of dikes and quays and south of the river dominated by ring polders
Source: Palmboom (1987)

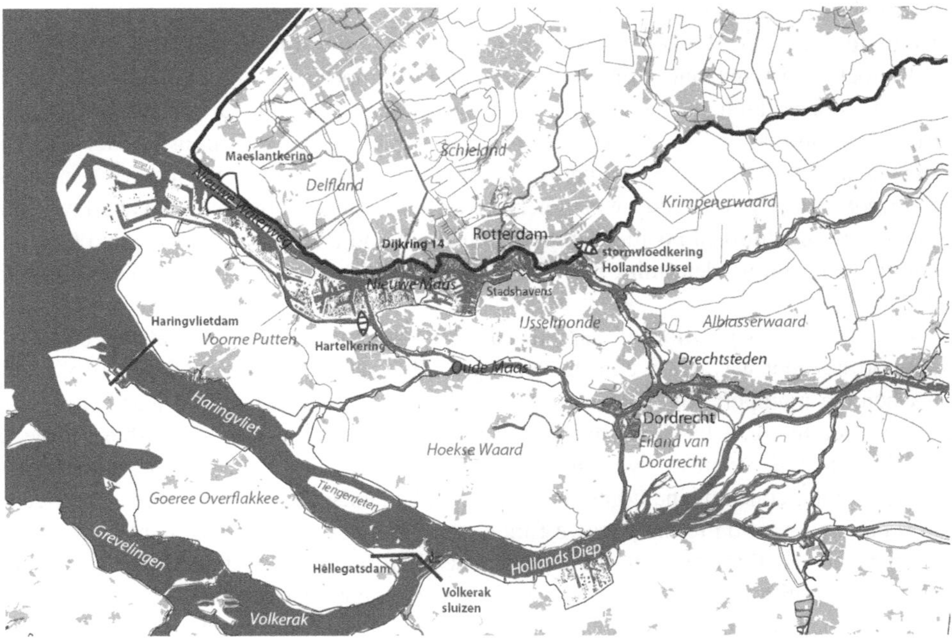

Figure 4. Rotterdam region ("Rijnmond") with unembanked and port areas
Source: Authors

vessels? Although the number of inland vessels entering the mainport is pretty impressive—about 110,000 in 2009 approximately—the modal split of the harbour results in a heavy environmental burden on the Rijnmond metropolitan region and the country as a whole. Although one should not overlook the fact that emissions from inland vessels are an important polluting factor, the overall majority of container transhipment does not take place via waterways but via roads. Of the 4132 million containers going to or coming from the hinterland, no less than 56% is transported via lorries, 33% via inland vessels and only 11% via rail.[2]

From the late 1960s onwards, the attention for environmental issues increased and the development of the Rotterdam mainport attracted the attention of environmentalists who started to voice opposition. In the late 1980s, the debate entered a new period because of plans to build a dedicated freight railway line from Rotterdam to Germany as well as (again) a new, large-scale extension of the harbour area: the Maasvlakte 2. Both projects would have serious environmental repercussions. Environmentalists were joined by—at least some—economists claiming that more development space and improved accessibility were an outdated strategy for strengthening a mainport. For instance, Boelens and Atzema (2006) claimed that instead of focusing on volumes of transported goods, a network-oriented strategy would make far more sense.

The concept of a mainport approach was born in Rotterdam as Van Duinen (2004, p. 64 ff) explains. Two professors of Erasmus University of Rotterdam—Poeth and Van Dongen—published several reports in the early 1980s on port development. They highlighted the global trend to concentrate on specific activities in huge, centrally located ports, a trend resulting from worldwide shipping processes. For the sake of efficiency, shipping companies would direct their ever bulkier vessels to a small number of ports: from "multi-ports" to "main ports".

This signal was rapidly picked up by the Rotterdam Port Authority but in doing so part of the story was overlooked. Instead of just concentrating on the qualities of the ports itself, Poeth and Van Dongen emphasized the crucial importance of looking at entire international transport chains: advanced logistical systems, hinterland connections, inland terminals, automation, etc. (Van Duinen, 2004, pp. 65–66). According to Van Duinen, this was a bridge too far. Rotterdam port authority stayed within its span of control: a (main)port is a physical entity which needs excellent external connections via all relevant modes. The institutional position of an organization and its internal doctrine and (spatial) planning conceptions determine how the world looks like (see, for instance, Throgmorton, 1992).

At first public opposition was small. An obvious explanation is that at this stage the mainport concept was just a concept on paper. The full effect only became visible later on when the mainport concept was used to substantiate plans to build Maasvlakte 2, the Betuwe freight route to Germany and also Schiphol's fifth runway. All these projects have been heavily contested with judicial fights in the highest courts in the Netherlands. Environmental concerns including the effects on the landscape—which were of particular importance in relation to the Betuwe Line—were at the top of the opponents' agenda.

Ultimately all three projects have gone ahead, although two of them are still under public and political discussion. Schiphol is the object of ongoing rows about noise pollution (on the whole only NGOs and the GreenLeft party emphasize the negative effects of air transport on CO_2 emissions). A main issue concerning the Betuwe route was the dramatic increase in the estimated costs. This eventually led to an official parliamentary investigation, including public hearings. Moreover, the line—which started to operate in 2007—is still not properly connected to the German railway system. The original plans foresaw a connection in 2003. The present indication from Germany is that this is not likely to happen before 2020.

It was several years after the birth of the mainport concept before the Port of Rotterdam developed a strategy which focused on transport "chains" instead of just on the port area itself. The port authorities started negotiations with inland ports in the region and the hinterland with the aim of a better control of the quality of hinterland connections which is of vital importance for the position of the port itself. This resulted in the management of the port of Dordrecht by the Port of Rotterdam from July 2011 onwards and plans to become a shareholder—together with Antwerp—in the port of Duisburg (Germany), the largest inland port in Europe. These latter plans have been heavily opposed by the Port of Hamburg though.[3] Also negotiations concerning collaboration and coordination with the ports of Amsterdam and Antwerp have recently started. What will come out of this is highly uncertain though: these three ports have always competed with each other. Moreover, the port authorities of Amsterdam and Antwerp are still municipal services while the Port of Rotterdam is privatized, its shares owned by Rotterdam municipality and the national government. In the case of Antwerp and Amsterdam, perceptions of territorial interest and the competition of municipal councils might still tip the balance.

3.2. *Maasvlakte 2: Sustainability Conflicts*

During the 1970s, it became clear that due to the changing port economy and transport technology the inner city port areas would be abandoned. The port concentrated its activity in the large-scale post-war port areas of Botlek, Europoort and Maasvlakte (Meyer, 1999). As a result, large sections of the inner city harbour areas became available for urban transformation. The second half of the 1980s turned out to be a period of economic recovery. Growth figures for the Rotterdam port were rising resulting in a new plan for the port: the 2010 Port Plan. The plan was officially adopted by the municipal council late in 1993 although there was some opposition because of environmental concerns. The main proposal was namely to create a new major port immediately west of the latest Maasvlakte 1 extension.

The role of environmental groups became powerful when the realization grew that the future development of the Port of Rotterdam not only falls under Dutch law, but also under European law: the 1992 EU Habitat Directive. In 2000, this directive had still not been properly transposed into Dutch law as required under the EU treaties. But that does not change the working of an EU directive: in the absence of a (proper) transposition, a European directive takes effect directly. This strengthened the position of environmentalists opposing the plan to build the new port area.

Eventually, a compromise was reached between the various actors and arenas. Of particular importance was an agreement between the municipality of Rotterdam and three large environmental groups laid down in a memorandum of understanding in 2000 called "Vision and Courage". In this document, both parties agreed that the objectives to further develop the Rotterdam mainport are of equal importance as the improvement of the liveability in the Rijnmond areas. A second Maasvlakte could be realized and compensated by various measures. Also several projects to develop new nature and outdoor recreation areas up to a total of 750 hectares had to be realized (this would eventually lead to four projects in Rotterdam). Within the existing port area, the available space would be more intensively used and measures would be taken to improve the environmental impact of industrial and harbour-related activities. Easterly port areas that were no longer needed would acquire new urban functions. At the time of writing, work has already started; their progress can be followed through a webcam.[4]

4. The Ecological Footprint of Rotterdam Mainport and Its Possible Future

4.1. *The RCI*

The debate about Maasvlakte 2 and the making and future use of Maasvlakte over the course of time have become increasingly framed within the perspective of climate change. As Rotterdam is a major transport hub as well as a (very) large concentration area for industry and especially the petrochemical industry, environmental policies are becoming increasingly intertwined with climate change mitigation policies and energy transition strategies. In both these areas, "Rotterdam"—that is, a coalition of various stakeholders—is striving to become world leading and a benchmark for many other port and delta cities.

The major vehicle for this ambition is the RCI in which energy transition and the reduction of CO_2 emissions play an important role in the aim to make the city "climate-proof" by 2025. The RCI is a cooperation body with four members: Rotterdam municipality, Port of Rotterdam, DCMR Environmental Protection Agency Rijnmond and Deltalinqs. The latter is a regional NGO which—according to its website—"[...]represents the common interests of all the logistical and industrial companies in the Rotterdam port and industrial area. The organization is considered to be the focal point and spokesman for more than 600 registered companies and associations."[5]

In public nearly always represented by its ambassador, former prime minister and UN's High Commissioner for Refugees Ruud Lubbers, RCI has a very simple slogan: 100% climate-proof. It seeks to create "[...] a movement in which government, organizations, companies, knowledge institutes, and citizens collaborate to achieve a fifty per cent reduction in CO_2 emissions, adapt to climate change, and promote the economy in the Rotterdam region."[6]

The RCI CO_2 reduction targets are ambitious when compared with those set by others. The region of Rotterdam for instance—a statutory cooperation body in which Rotterdam municipality participates—is seeking a reduction of 40%. The province of South Holland is following general EU policy targets: a reduction of 20% (OBR, 2010, p. 97). This also counts for the present centre-right national government which explicitly does not strive to move beyond the ambitions of the EU. What is more interesting is how the government would like to reduce CO_2 emissions, namely through large-scale carbon capture and storage (CCS) and nuclear energy. The latter is unprecedented when compared with a number of previous government coalitions: the present government explicitly sees nuclear energy as a climate friendly technology.

CCS is a main component of the RCI strategy. CCS has become urgent because there are plans to build two new coal-fired power plants in the port area. Here, the CCS strategy of the RCI received a major blow in November 2010. A major project to store CO_2 in a depleted gasfield under the city of Barendrecht has officially been abandoned after heavy protests by the local community, afraid of certain risks such as leaks from the installations or deep from underneath the soil. The North Sea is full in the picture now. Plans to store CO_2 in the continental shelf are already being carried out. As is the case with offshore wind farms: potential opponents are literally over the horizon.

4.2. *To Maintain Mainport Status in an Age of Energy Transition*

We can say that port areas at present are indeed hubs in the global and continental flows of energy. In this respect, no other European port has the same status as Rotterdam. To maintain this position within the perspective of energy transition is part of the survival strategy of the Port of Rotterdam and other major stakeholders involved in the port area.

Being an energy hub is at present almost exclusively related to fossil fuels: oil, coal and—in the nearby future—liquefied natural gas (LNG). The throughput of oil at the port is pretty stable at the moment at around 100 million metric tons annually. About half is transported through pipelines to Vlissingen in the southwest of the country and Antwerp and Germany. The other half is processed in the port area itself: together with Singapore and Houston, Rotterdam belongs to the global top three of petrochemical clusters. Throughput of coal has steadily risen over the years and is now about 25 million metric tons. Transported by belt (in the Rotterdam port itself), barge and rail about 20% is used in steel mills, the remaining part in power plants mainly in Germany.[7] A main goal is that a third energy flow will be added to the classic two formed by oil and coal: LNG. This is part of a wider, national strategy to diversify the importation of natural gas to improve the security of the supply. But Rotterdam is also striving to play a major role in the European transhipment of LNG as well as building an industrial cluster around LNG in the port area itself.

At present the port is also developing a strategy to become a hub in the transhipment and use of biofuels and biomass. With this in mind the so-called Rotterdam Biomass Commodities Network (RBCN) has been set up, an offshoot of RCI and formed by a large number of companies.[8] As biofuels have come under heavy attack by environmentalists and NGOs—through the cutting of (rain)forests and pushing aside food production—RBCN supports the certification of biofuels mainly through the Dutch NTA8080 standard.[9] As with other fuels, biofuels are supposed to play a role in energy production within the port area itself (there are plans to build several energy plants) as well as becoming the core of industrial clusters. This is called co-siting by the Port of Rotterdam Authority, whereby new facilities link up directly with adjacent tank terminals and factories.

What the effects could be of the future energy transition—the general heading for the period during which, according to many, fossil fuels will be gradually phased out—for Rotterdam is the object of discussion and scenario building. At the time of writing—early Spring 2011—the port authority is preparing its new 10-year vision, its horizon 2030. The full effects of energy transition will be felt later according to many. At present the harbour is still developing in a mainstream fashion as the above has pointed out: for the years to come, fossil fuels will continue to form the foundation of the harbour, their negative effects being overcome by CCS. Although the overall objective is to maintain the position of being an energy hub on the continent, also during and after a possible period of energy transition, whether this will indeed be the case is open to speculation. There are many unknown factors and quantities. Bio fuels form an entirely different sort of commodity compared with oil and coal. In particular, the concentration of production and transport of these classic fuels is totally different when compared with biofuels. Rotterdam's position in the global top three of petrochemical clusters is largely due to its location: in the centre of a mega-region, connected to a major mass-transport axis (the Rhine) and accessible for the largest vessels on the planet. If the future pattern of biofuel and (sustainable) energy production is to become much more deconcentrated, which seems to be becoming the case, transport patterns could become totally different. More spatial inertia could stem from the spatial pattern of chemical complexes. If the Rotterdam petrochemical complex is to gradually turn into a biochemical complex, the sheer investments needed would lead to some form of stability in the transport chains of biofuels and the biological replacements of (crude) oil and their (semi)manufactured product.

5. Sustainability of the Delta City

Besides energy transition and the reduction of CO_2 emissions, water and water management play an important and multiple role in the RCI programme. This means that policies concerning sustainability necessarily need to take into account the different aspects of water in the city. These are

(1) groundwater control and soil subsidence;
(2) rainstorm management;
(3) rises in sea level;
(4) increasing river discharges;
(5) environmental quality and bio-diversity;
(6) salinization.

5.1. *Groundwater Control and Soil Subsidence*

Like many other delta cities, Rotterdam is dealing with a difficult dilemma concerning groundwater control. Originally the peat areas were very wet and unsuitable for human living and building. Draining these areas had already started in the twelfth century and continued over the following centuries, supported by improving technology. A result of draining peat, however, is that the dried peat starts to oxidize which results in shrinkage. As a consequence, the ground level subsides by about 1 m each century, resulting in ground levels varying from 1 to 6 m below average sea level (referred to as Normaal Amsterdams Peil (NAP)). This subsidence means that the consequences of flooding if a dike broke would be disastrous. Taking into account a storm surge with a sea level of 4 m above NAP, the result of breaking dikes would be that the water level in the flooded city would be 5–10 m above ground level.

Stopping the process of soil subsidence is crucial in order to make the city less vulnerable to flooding. It means that the groundwater level needs to be maintained as high as possible, which conflicts (especially in the older parts of the city) with the interests of people owning houses with basements which are vulnerable to high groundwater.

5.2. *Rainstorm Management*

Because of climate change, northwest Europe is confronted with an increasing frequency of heavy rainstorms, delivering much more water in a short time than was usual. During the last 100 years, the average rainfall has been 790 mm per year, but this is increasing rapidly. Estimations calculate the increase to be 20%, while the intensity of rainstorms is also increasing (Gemeente Rotterdam *et al.*, 2007).

During the nineteenth and twentieth centuries, a lot of the original canals in the city of Rotterdam were filled. They not only lost their function as transport routes but new drainage technology focused on the construction of underground draining systems. As a result, the amount of surface water decreased from originally 16% to less than 6% of the built-up area. Because of this decrease, sewage systems are frequently over-loaded during periods of heavy rainfall. Pumping stations are not able to pump the water from the sewage system into open water (river or sea) fast enough, resulting in flooded streets and the spillover of sewage water in what is left of the canal and open water system within the city.

Instead of increasing the capacity of the sewage systems and pumping stations, the focus of water management in Rotterdam (and other Dutch cities) has shifted to increasing the storage capacity for water in the city. This means making new surface water areas as well as the creation of public spaces which can play a role for the temporary storage of water. This policy is linked to a more sustainable groundwater management: an increased amount of surface water supports a higher groundwater level (Gemeente Rotterdam *et al.*, 2007).

5.3. *Rising Sea Levels and Increasing River Discharges*

Rising sea levels and increasing river discharges are two different things. However, the combined effect creates new problems for the city. Rising sea levels is not something new but has already been taking place for thousands of years. During the twentieth century, the sea level rose by ca. 17 cm. The "speed" of the rise in sea level, however, is supposed to increase due to climate change. The various calculations and estimations of, for instance, the Intergovernmental Panel on Climate Change do not give any certainty about the exact situation in the future. The so-called Second Delta Committee installed by national government—the first Delta Committee was installed after the disastrous 1953 flood which killed about 2000 people—assumes a considerable increase of the sea level during the decades to come. The Delta Committee bases its advice on the maximum scenario of a 130 cm rise in sea level by the end of this century (Deltacommissie, 2008). Apart from the question of whether this scenario is realistic, it is generally assumed that a structural rise in the sea level will take place.

At present the dike system runs right across the urban fabric of Rotterdam. The unembanked areas, directly adjacent to the river, have been raised to a level of 3.20–4.00 m above NAP in the course of the nineteenth and twentieth centuries. These levels were supposed to be sufficient to avoid regular flooding, but they will become increasingly vulnerable to flooding in the future. To prevent flooding, a storm barrier in the mouth of Nieuwe Maas has been built: the Maeslant Storm Surge Barrier.[10] It closes when the sea level is expected to rise by more than 3 m above NAP. Without this barrier, a large part of the areas outside the dikes would be flooded by several centimetres during serious storm surges.

With the expected rise in sea levels, it is expected that the Maeslant barrier will have to close more frequently in the future. This will produce substantial problems for the port. The present frequency of closing the barrier is about once every 10 years which is regarded as acceptable for the continuity of the port economy. However, a substantial increase of the frequency of closing, combined with the uncertainty of when the barrier will be closed, is regarded as an unacceptable decrease in the reliability of the open entrance. We will come back to this later.

Increasing river discharges have been a reality since the mid-1990s, when the river area in the Netherlands was confronted several times with either serious floods or near-floods. According to the Delta Committee, river discharges could increase by 150% over the coming decades. Extreme river discharges of the Rhine, entering the Netherlands at the east border, deliver 12,000 m^3 per second at the present time, but are expected to increase to 18,000 m^3 per second in the future (DeltaCommissie, 2008). For the Rotterdam region, this means that the flood hazards come from two directions: from the sea as well as the rivers. A worst case scenario would be a coincidence of an extreme storm surge and extreme river discharges.

In the report of the Delta Committee, the Rotterdam region is appointed as a special case, which makes extraordinary solutions necessary. In 2010, a high-level Delta Commissioner was appointed by the government, combined with the establishment of a special programme committee for the Rotterdam region. Supported by several research initiatives,[11] this committee is investigating different scenarios and options for increasing the safety, varying from raising the existing dikes to closing the whole region by a system of locks or storm surge barriers and directing the river water to the delta south of Rotterdam.

Technically all these different options are possible; the big question concerns the effects of each type of solution on the port economy, the city structure and the environment. An important issue concerning city structure is that the dike system runs right across the urban fabric of Rotterdam. As already mentioned, the unembanked areas were raised to a level of 3.20–4.00 m above NAP in the course of the nineteenth and twentieth centuries. The ground levels behind the dikes are much lower, due to subsidence because of continuous drainage, as explained above. So a usual situation has occurred where the area "outside the dike" is much higher than the area inside the dike. In between, the dike has an average height of 5.5 m and is experienced as a barrier between the floodplain areas and the areas behind the dikes.

5.4. *Environmental Quality and Bio-diversity*

Since the 1980s, interest in the environmental value of estuaries has increased substantially, in particular because of their function in the complex ecosystems of connecting the seas, oceans and rivers (see, for instance, Saeijs, 2006). Several (Dutch) environmental groups, large NGOs (World Wildlife Fund, 2008), landscape architects (Sijmons, 2002; Sijmons & Venema, 1998), scientists (Tjallingii, 1996) and also governmental institutions are paying increasing attention to the environmental effects of large-scale hydraulic works in the southwest of the country: damming the estuaries resulted in dramatic changes to their bio-diversity and in a concentration of pollutants, river sediments and agriculture toxins (see, for instance, Adriaanse & Blauw, 2007). The title "Working together with water" of the report of Delta Committee is an expression of the recognition of this attention to environmental issues in official flood-control politics. It has resulted in a serious reconsideration of the future position of the Delta Works, the series of dams and storm surge barriers in the delta in the southwest of the country. NGOs like the Word Wildlife Fund put a lot of energy into their pleas to maintain or restore the openness of the large estuaries worldwide. They have chosen the Dutch delta as an important test case and developed a plan for this delta, showing a complete reopening of all the estuaries (WWF, 2008).

For the Rotterdam region, this means that also here the combination of flood defence, port installations and environmental issues has been placed high on the political agenda. The clearest example so far is (as described in Section 3) the layout of a series of new natural areas in the Rotterdam region as a compensation for the loss of natural environment because of the construction the new Maasvlakte 2. The construction of this new reclaimed land in the sea will change the environmental system of the coast; the new natural areas are supposed to compensate this loss to a greater extent than required by the EU Habitat Directive (see below). However, this solution can be regarded as a rather artificial and incidental approach; for future challenges, a more comprehensive approach will be expected.

5.5. *Salinization*

The changing climate results not only in more rainfall and intensified rainstorms, but also in longer periods of drought. Combined with rising sea levels, this phenomenon leads to a shortage of fresh water and an increasing influence of salt water in the peat and clay areas of Holland. This process has a damaging effect on the agriculture in the Rotterdam region. Due to the Delta Works, the estuaries Haringvliet and Brielse Meer have been transformed into large suppliers of fresh water. This guaranteed fresh water supply on a large scale (a unique condition for a sea port) was a main motive for the petrochemical industries to settle in Rotterdam in the 1960s, resulting in the second largest petrochemical complex in the world. Today, the port–industrial complex uses more than 36 million m^3 of fresh water each year (Stuurgroep Zuidwestelijke Delta, 2009).

A supply of fresh water is also essential for agriculture. The clay polders south of Rotterdam have become one of the most productive agricultural areas in Europe (Ruimtelijk Planbureau, 2005) and the "Westland" area just northwest of Rotterdam has been developed into the most intensive complex of greenhouses in Europe. This agri- and horticultural area is also dependent on the availability of fresh water on a large scale and is connected with a pipeline to the Brielsemeer.

6. Promises and Problems of New Urban Waterfronts

The recently increased interest in the effects of climate change on water management and flood defence, and the increased attention to competitiveness and identity have created perspectives and chances for a new interweaving of targets concerning hydraulic engineering and targets concerning urban and economic regeneration.

As a result of its historic development, Rotterdam is a city with a strong difference between both sides of the river: Rotterdam Centre and Rotterdam South. Both sides are different in the sense of spatial structures, economic activities and social and cultural structures. Currently, the two parts of the city have different safety standards concerning flooding. The northern dike-ring 14 area (Figure 1) is provided with the highest safety standard, with a chance of flooding of once in 10,000 years. Rotterdam South is part of a dike ring with a different chance of flooding: once in 4000 years.

With the departure of port-related industries during the 1980s and 1990s, Rotterdam South became one of the most problematic urban areas of the Netherlands, characterized by high rates of unemployment, poverty, low degrees of education, racial tensions, etc. Since the 1980s, the city administration has tried to change the one-sided focus of the economy on the port to a more diverse urban economy. Approximately 15,000 residential units, mixed with offices and amenities, have been built on approximately 250 hectares of former port areas, like, for instance, in the Kop van Zuid area at the left bank of the river. All of these developments took place in unembanked areas.

During the planning and construction of the first generation of transformation projects, the city of Rotterdam agreed upon a provisional rule with the provincial and national authorities: the entrances and floors of new houses in unembanked areas should be raised to a level of 4 m above NAP. This rule was regarded as sufficient, taking into account the new "Maeslant" Storm Surge Barrier that was constructed close to the mouth of the Nieuwe Waterweg during the 1990s and that is intended to close whenever the sea level rises more than 3 m above NAP, as explained in the previous section. However, the growing danger of high water levels because of increasing peak discharges of the rivers that feed the delta, as also described in the Section 5, has forced the city to develop new strategies.

The need for a comprehensive strategy is becoming increasingly more important, as a huge unembanked port area will shortly become available for urban (re)development: the so-called city ports area that covers 1000 hectares. The city ports provide great potential in terms of new urban districts in attractive environments and creating more spatial coherence between the two parts of the city on both river banks (Meyer 1999); at the same time, they are part of the most flood-prone areas in the urban territory. One of the main aims of Rotterdam's urban policy will be solving this paradox.

6.1. *Perspectives for Solutions*

In general, there are three basic principles that can be used to protect the unembanked areas against flood damage (De Hoog & Nillesen, 2010). The first one is to dam the water off by creating dikes, barriers or dams that keep the water out. This can happen on a regional scale (see next section), but also on a local scale, at the site of a building. This method, however, creates a similar barrier between the built-up areas and the water as is nowadays found to be problematic in Rotterdam's city centre. There are new options for designing dikes in such a way that they become part of the waterfront, by integrating urban space, nature and built functions within the body of the dike (Stalenberg, 2010).

A second option for making the unembanked area flood-proof would be to elevate either the whole area or the building grounds. This method is one of the oldest water defence mechanisms used in the Netherlands. Farmers from the northern provinces built their houses on artificial mounds called "terpen", to protect themselves from floods. Elevation is frequently used in the Rotterdam harbours as well when creating harbour sites, such as, for instance, the Maasvlakte. It is the same principle that is successfully used in the Hamburg Hafencity, an old harbour in Germany that is being transformed into a flood-proof residential and business area. In the former port area, a partial elevation has been created. Next to the elevation, buildings are placed on flood-proof plinths that at the same time serve as parking garages. In case of high water levels in the river Elbe, the garages are locked and the lower parts of the streets, the original quays, are flooded. A secondary network of pedestrian evacuation roads connects the buildings with the elevated main street. In the Hafencity project, building a dike around the building site was considered but was also rejected as the relation with the water would have been compromised and a large upfront investment in water safety would have been needed. By making each building block flood-proof separately, the investment in flood protection grew in line with the pace of development, reducing the economic risk in case of stagnating development. The public space is brought close to the water and designed with multiple layers in order to optimally experience the fluctuation and natural force of the water.

The third principle is to make buildings flood-proof by, for instance, placing them on poles, using floating or amphibious houses or by making the interior of houses waterproof. The latter options are still in an experimental stage and in urban developments often considered as small-scale solutions rather than as overall urban flood strategies. Floating houses and functions are very popular with designers, even though they occupy water instead of land. Therefore, the amount of floating functions the water can hold without losing its openness—an invaluable quality the river offers the city of Rotterdam—should be studied carefully.

Since the Rotterdam city ports aim to be an experimental site for waterproof innovations, a wide range of flood-proof measures for buildings and public spaces could be applied within the area, creating new urban typologies that deal with the water in ways that are distinctly different from existing practices. One of the essential questions will be how to create well-working connections between the harbour areas being transformed and the rest of the city. Apart from the existing industrial transportation network, the sites are poorly connected with their surroundings and the city centre. Socially and economically poor neighbourhoods, as well as a large amount of infrastructure, form a barrier that creates a disconnection, especially between the southern harbour sites and the city centre. Transportation over the water is now intended to better connect the transformation areas.

7. Flood Protection: Crux of a Comprehensive Strategy

The Delta Programme proposes four flood-risk strategies that define the corners of a "playing field" of possible safety solutions for the Rijnmond region. Many different aspects of sustainable development in Rotterdam (port development, safety against flooding, possibilities for new forms of urbanization in the former dockland and waterfront areas, and care for environmental quality of the delta) are strongly interlinked with the choice for a regional flood defence strategy.

The four strategies are based on the principle that a storm surge barrier reduces the amount of inland dike reinforcements, by closing the front entrance towards the sea and thus protecting the region against high water threats. The same basic principle led to the construction of the existing Maeslant Storm Surge Barrier that currently protects the Rijnmond region against high water levels in the case of a storm surge.

The first water safety strategy is based on a "business as usual" approach, protecting the Rijnmond region with the existing, closeable Maeslant barrier at the sea-bound side of the region (Figure 5). Barriers which can be closed carry with them a chance of failure, however, meaning that the dikes behind the barriers have to be strong enough to protect against rising water levels in the case of barrier failure, or in the case of high river discharges (that are expected to become more extreme and increase from 12,000 to 18,000 m^3 per second at peak discharge) occurring simultaneously with a storm surge. Due to rising sea levels, this might (as mentioned in Section 5) in the future result in

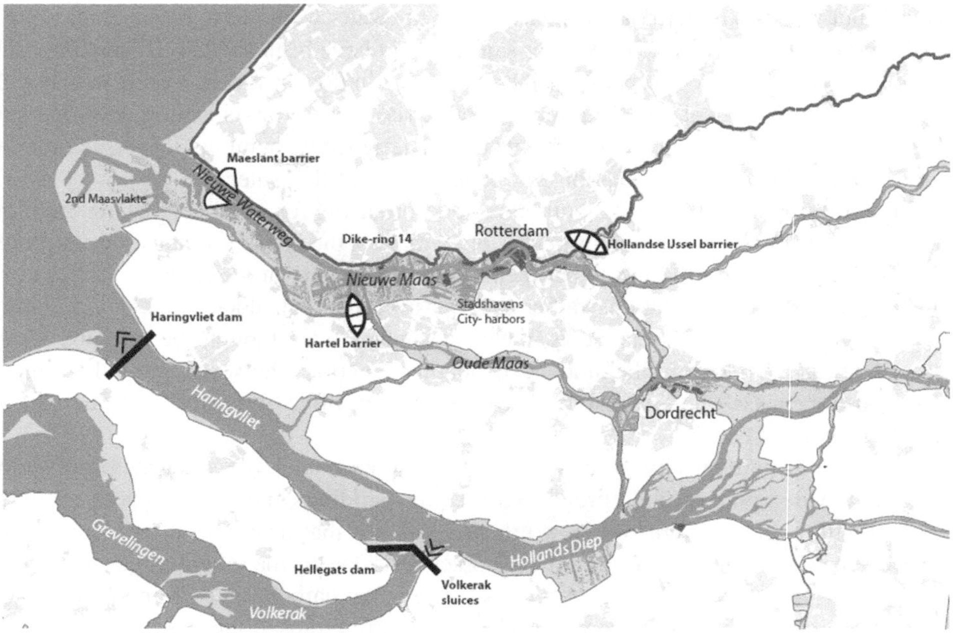

Figure 5. Rijnmond open/closable at seaside, with storm surge barriers in the sea gates
Source: Authors

more frequent closure of the barrier. In an extreme scenario put forward by the Delta Committee (Deltacommissie, 2008), the sea water level might have risen 1.30 m by 2100. The barrier would then have to close up to 30 times a year—unacceptable for the Rotterdam port authority, since the accessibility, and with that the reliability, of the port depend on the open connection with the sea. Besides that, the chance of the Maeslant barrier failing to close is estimated to be one in every 100 closures. More frequent closing of the barrier would lead to a greater risk of failure in the future. Therefore, the Delta Programme proposes (Deltaprogramma Rijnmond Drechtsteden, 2010) that the barrier, when it comes to being replaced, should be upgraded to a barrier with an improved chance of failure of 1 in every 1000 closures. It is also proposed that the barrier should not be closed more frequently than once a year, even if this results in closing the barrier at a higher water level than the currently prescribed +3 m NAP. With some of the unembanked areas starting at a height of 3.2 m + NAP, this means that additional measures would be needed to protect these areas against flooding.

The second water safety strategy is an answer to the extra flood risk caused by a simultaneous storm surge and peak river discharges by placing additional closeable barriers on the river side of the region (Figure 6). In case of peak discharges, the barriers could then be closed and the water redirected towards the Haringvliet that directs it towards the sea. An extra advantage of the placement of barriers is the possibility to use them to improve the infrastructural network of the region. However, the small-scale "business as usual" interventions in between the barriers that are related to this strategy hardly create a catalyst for new urban transformations of a scale that could redefine the relation between the north and southern part of Rotterdam.

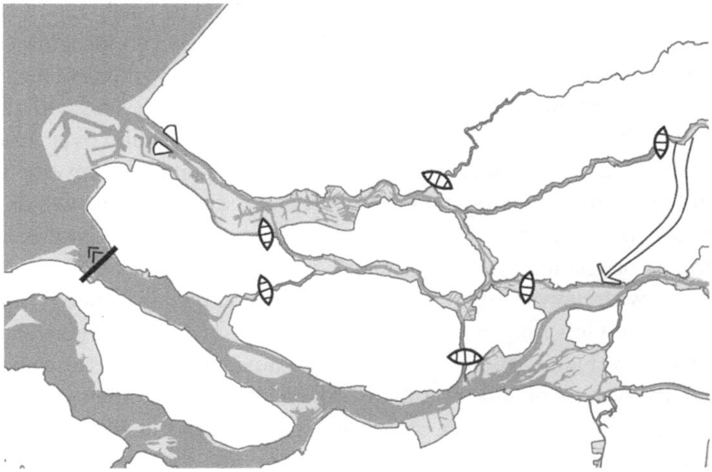

Figure 6. Rijnmond open/closable at seaside and riversides—with storm surge barriers in all waterways to the urban territory
Source: Authors

The third strategy, "closed Rijnmond", provides a solution to the problem of the chance of failure related to the closeable barriers by damming off part of the Rijnmond region—leading to a similar situation as in Amsterdam (Figure 7). The new barrier would disturb the natural, open connection of the river towards the sea. The Port of Rotterdam would lose its big competitive advantage: open access to both the North Sea and its European hinterland. Ships would have to pass locks to access the port or use hinterland connections. However, in between the barriers a controlled water level would create many possibilities for urban redevelopment and would allow a transformation of the central river part of the city, which now divides the city into two parts. Because of the controlled water level, the

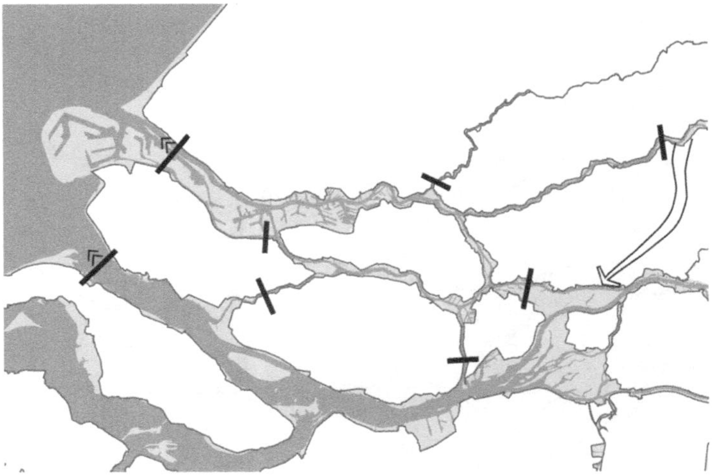

Figure 7. Closed Rijnmond, with locks and dams in all waterways to the urban territory
Source: Authors

dikes forming a barrier between the city and the river could be partly removed or lowered in order to strengthen the relation of the city with the water. By gaining a more recreational and urban function, water could become part of the city itself and connect the two halves of the city. The closed system makes it possible to connect the river Maas with the canals and creeks within the dike rings, creating possibilities for transportation over water as well as ecological connections. In this scenario, the river has the potential to develop into a central spine. Depending on the development of the port and the need for growth or shrinking of the city, new urban environments could be created that include mixed urban areas, port-related functions, or green recreation or nature areas.

The fourth safety strategy is referred to as open Rijnmond. In this strategy, all storm surge barriers including the Grevelingendam would be removed, creating an open relation between the rivers and the sea (Figure 8). The natural tide, sedimentation processes and the fresh and salt gradient of the water would be re-established, resulting in a more natural estuarine delta area. For the shipping traffic, this open connection guarantees accessibility of the port and the hinterland. The unembanked areas would come under tidal influence and flood on a more regular basis. Any areas containing buildings, polluted grounds or industries vulnerable to floods would have to be protected by additional small-scale dikes or barriers. The regeneration projects within the unembanked area can be designed as flood-proof areas, strengthening the relation between the river and the city. To reach the safety levels prescribed for the dike rings, most dikes would have to be reinforced in a drastic way. New forms of Delta dikes could be used to strengthen the dike while at the same time making the dike itself become part of the urban waterfront. From this perspective, the open Rijnmond strategy offers changes that are comparable to those of the closed Rijnmond strategy—to give the river a more central and binding position within the city of Rotterdam.

The range of possible water safety strategies defined by these four "corner positions" creates possibilities to compare them and distinguish their characteristics, effects and opportunities. The four options ("corner positions") result in different conditions for water levels in the unembanked areas, as shown in Figure 9 and with that in different conditions and possibilities for the urban environment.

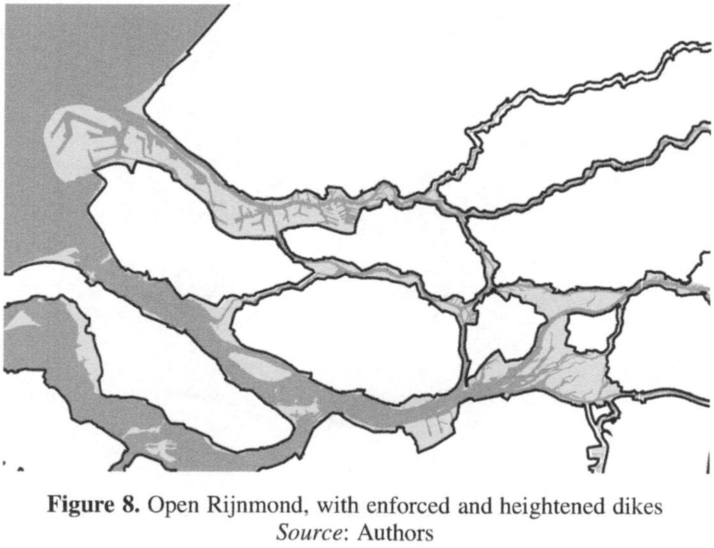

Figure 8. Open Rijnmond, with enforced and heightened dikes
Source: Authors

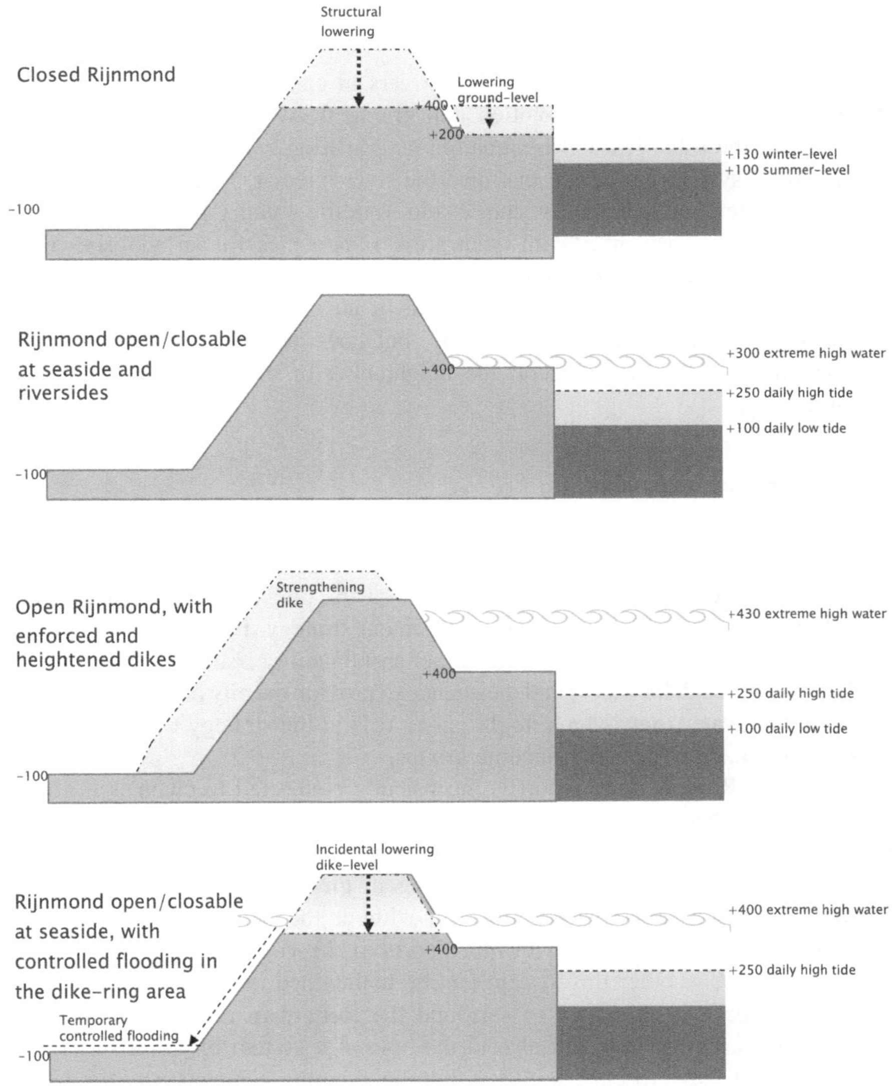

Figure 9. Dike sections showing the effects of the different regional water management options for the local situations of unembanked areas and dikes
Source: Authors

Also the future of the port will be influenced seriously by the type of flood-protection strategy. However, the interests of the port are not unambiguous. Considered from the perspective of the importance of accessibility of the port area, the option of an "open Rijnmond" seems to be the best one. But considering the importance of the availability of fresh water supply for the process industry, the closed option delivers the best conditions. And considering the long-term perspectives, with a possible transformation from petrochemical industry into other types of production and storage of energy, the question is if the port will need the total amount of the present-day 10,500 hectares of port area in the future. The answer to this question will be important in the case of a decision concerning the location of a new storm surge barrier or lock system.

Finally, the environmental quality of the rivers in and around the Rotterdam region will be dependent on the final solution concerning flood protection. An open system will mean that the estuaries will be repaired as transition zones between salt and fresh water; a closed system will mean that also the river Nieuwe Maas will be transformed into a fresh water body. Recently, the World Wildlife Fund (WWF, 2010) presented its report "Met open armen" ("with open arms"), as a plea for an abolishment of the dams in the southwest delta and for the repair of estuarine wetland systems in the river mouths. The WWF states that the repair of an open system will contribute to a new balance of nature, safety and economy, but does not say what the consequences would be for safety in this concept: the heightening of many kilometres of dikes in the region.

8. Conclusion and Discussion

The Rotterdam region plays a central role in two main issues concerning sustainability in the Netherlands: water management and energy transition. These two issues are directly linked to each other because of the central role of the port as a global energy hub and as a central element in any new water management strategy. Furthermore, the decisions concerning these issues directly influence other sustainability issues, such as the environmental quality of the delta waters and the chance to provide the city of Rotterdam with new attractive urban waterfronts which might play a role in the strategy to attract innovative industries and medium- and high-income groups.

The future of Rotterdam mainport is intrinsically connected to changes in the Dutch delta system and therefore to one of the main effects of climate change: the rise in the sea level. The first major step on the road to the present status of being a global mainport—the realization of the Nieuwe Waterweg in the nineteenth century—created an open port, accessible directly from the sea without the necessity of passing a lock barrier, an advantage compared with other ports in the Hamburg—Le Havre range. It remains to be seen whether this system can be maintained in the long run. Worst case scenarios will make it necessary to surround the Rotterdam region with a system of locks or storm surge barriers. But also in the case of a system of storm surge barriers, these barriers will have to close so often that the shipping connection to the hinterland will become unreliable and "unacceptable" financial damage to the shipping sector will be the result.

The paper shows that altogether, "water" means several different issues, which touch and overlap each other. Moreover, many of these issues touch spatial, economic, social and environmental questions in the city and harbour area. The question is how a planning approach and policy can be developed which considers the linkages between the different issues of water and the linkages of water with spatial, economic and social issues as a possibility to develop comprehensive plans.

Summarizing, to answer the question of whether Rotterdam really is a worldwide benchmark concerning water and port management: perhaps yes in the sense of ambitions and intentions; in the sense of real practice, there is still a long way to go.

On the positive side, it is clear that many initiatives are taking place at the moment. This paper has discussed several of them. The main challenge though is to link the various individual initiatives, strategies, studies, projects and the actors and coalitions behind them. There is, for instance, the process which should lead to a new strategy for the port (Port Vision 2030) and which—among other things—will deal with an issue totally new to the port authority: climate change. It does not yet seem to be the case that thinking about the future of the port has been linked to the issue of a long-term hazard-proof system of flood control. How water management could be linked to urban renewal and the revitalization of Rotterdam South is yet another example of how linkages can be made between issues but which is still in its infancy.

At first sight, this seems to be a call for comprehensive planning. This is not what we mean. Comprehensive planning presumes—among other things—the presence of decisive actors. These are simply not there, have never been there nor will ever be there. What we would like to see is the making of linkages. One way of achieving this is to give a key role to the water system because this system is so closely linked to the way the city and the port is developing and could or even should develop in the future. One example of this close relationship we have already highlighted above: the dependence of the petrochemical complex of Rotterdam—as one of the largest in the world—on a reliable and large supply of fresh water.

A water system is a territorial network in itself and based on these characteristics networks can be developed at the level of strategies, programmes, projects and actors. A serious handicap for such a network approach is the extremely complex government system in the Rotterdam region and the south wing of the Randstad at large. What is needed is what Bob Jessop call metagovernance. But it is very difficult to see which actor(s) could take the lead to create some kind of metagovernance capacity. Some kind of administrative restructuring appears to be an obvious strategy. There is doubt in the literature (see, for instance, Salet, 2006) whether this would be the right way forward, especially in the case of the Randstad and its sub-regions like the north (centred around Amsterdam) and the south wing (centred around Rotterdam and the Hague): the territorial organization of government will never perfectly match territorial relationships. Moreover, every new design of this organization is likely to become outdated in the future thanks to processes of spatial rescaling. Nevertheless, one could or should ask at what point a certain territorial arrangement becomes unworkable. This seems to be the case in the Rotterdam region. Starting to discuss the territorial structure of government in the Netherlands is like opening Pandora's box as has been proven by decades of discussion. At present there is yet again talk of rearranging the governmental structure of the Randstad. A genuine political pandemonium is the result: provinces, municipalities and their umbrella organizations take different positions, as has always been the case. But some administrative reorganization seems necessary: the use-by date of the present organization passed a long time ago.

Notes

1. www.rotterdamclimateinitiative.nl (accessed February 2011).
2. Of the 5.9 million containers transported via the Rotterdam harbour nearly 30% is called sea–sea transport (Port of Rotterdam, 2009).
3. http://www.rnw.nl/english/bulletin/germans-reject-dutch-purchase-port-duisburg-share (accessed February 2011).
4. http://www.maasvlakte2.com/nl/index/ (accessed February 2011).

5. http://www.deltalinqs.nl/index.php?option=com_content&task=view&id=54&Itemid=4 (accessed 8 February 2011).
6. http://www.rotterdamclimateinitiative.nl/en/about_rotterdam_climate_initiative/rotterdam_climate_initiative/mission_ambition (accessed 7 February 2011).
7. All figures collected from the website of the Port of Rotterdam.
8. Biomass is a wider concept than biofuels. Biomass is of particular importance for the Netherlands: according to some (for instance, Hetsen & Hidding, 1991), the large-scale very intensive cattle breeding which forms a major pillar under the Dutch agro-business industry owes its existence mainly to cheap and reliable imports of biofood through the Rotterdam port.
9. NTA is the Dutch acronym for Dutch technical agreement (see http://www.sustainable-biomass.org; accessed February 2011).
10. See http://www.keringhuis.nl/engels/home_noflash.html (accessed February 2011).
11. Especially the programme 'Knowledge for Climate', supported by NWO, the Dutch research council (see http://knowledgeforclimate.climateresearchnetherlands.nl; accessed October 2010).

References

Adriaanse, L. & Blauw, T. (2007) Naar een nieuwe delta [Towards a New Delta], in: K. Bos (Ed.) *Zeelandboek 11*, pp. 20–39 (Middelburg: Stichting Zeelandboek).

Boelens, L. & Atzema, O. (2006) Connecting Randstad Holland: A network approach of competition between clusters, *Nederlandse Geografische Studies*, 339, pp. 61–72.

Bouman, P. J. & Bouman, H. J. (1952) *De Groei van de grote Werkstad: Een studie over de bevolking van Rotterdam* [The Growth of the Big Work Town: A study on the population of Rotterdam], Assen: van Gorcum.

De Hoog, M. & Nillesen, A. (Eds) (2010) *Stedenbouw & Multifunctionele waterkeringen: Projectgroep Afsluitbaar Open Rijnmond* [Urban Design and Multifunctional Water Barriers: Project-team Closable Open Rijnmond], Delft: TU-Delft.

Deltacommissie (2008) *Samen werken met water: Bevindingen van de Deltacommissie 2008* [Working Together with Water: Findings of the Delta-committee 2008], The Hague. Available at http://www.deltacommissie.com/doc/2008-09-03%20Advies%20Deltacommissie.pdf (accessed 1 December 2011).

Deltaprogramma Rijnmond Drechtsteden (2010) *Plan van Aanpak* [Plan of Approach], Den Haag: Ministerie van Verkeer en Waterstaat, Ministerie van Landbouw, Natuur en Voedselkwaliteit, Ministerie van Volkshuisvesting, Ruimtelijke Ordening en Milieubeheer.

De Vries, J. & Van De Woude, A. (1997) *The First Modern Economy; Success, Failure, and Perseverance of the Dutch Economy, 1500–1815* (Cambridge: Cambridge University Press).

Gemeente Rotterdam, Waterschap Hollandse Delta, Hoogheemraadschap van Schieland en de Krimpenerwaard, Hoogheemraadschap van Delfland [City of Rotterdam, Waterboard Dutch Delta, Waterboard Schieland and Krimpenerwaard, Waterboard Delfland] (2007) *Waterplan 2 Rotterdam: Werken aan water voor een aantrekkelijke stad* [Waterplan 2 Rotterdam: Working with Water Towards an Attractive City], Rotterdam. Available at http://www.rotterdam.nl/waterplan_2_en_deelgemeentelijke_waterplannen (accessed 1 December 2011).

Hetsen, H. & Hidding, M. (1991) *Landbouw en ruimtelijke organisatie van Nederland: Analyse en toekomstverkenning van een regionaal gedifferentieerde betrekking* [Agriculture and the Territorial Organisation of the Netherlands: Analysis and future-reconnaissance of a regionally differentiated population] Wageningse Ruimtelijke Studies 6, Wageningen: Landbouwuniversiteit Wageningen.

Meyer, H. (1999) *City and Port; Urban Planning as a Cultural Venture in London, Barcelona, New York, and Rotterdam: Changing Relations Between Public Space and Large-Scale Infrastructure* (Utrecht: International Books).

OBR (Ontwikkelingsbedrijf Rotterdam) (2010) *Economische Verkenning Rotterdam 2009–2010—DEEL 1 Energie en milieu* [Economic Reconnaisance 2009–2010 Part 1: Energy and Environment] (Rotterdam: OBR).

Palmboom, F. (1987) *Rotterdam Verstedelijkt Landschap* [Rotterdam as an Urbanised Landscape] (Rotterdam: 010 Publishers).

Port of Rotterdam (2009) *Haven in cijfers* [Port Figures] (Rotterdam: Port of Rotterdam).

Ruimtelijk Planbureau (2005) *Waar de landbouw verdwijnt; Het Nederlandse cultuurland in beweging* [Where Agriculture Disappears: The Dutch Cultural Landscape in Change] (Rotterdam: NAi uitgevers).

Saeijs, H. L. F. (2006) *Turning the Tide: Essays on Dutch Ways with Water* (Delft: VSSD).

Salet, W. (2006) Rescaling territorial governance in the Randstad Holland: The responsiveness of spatial and institutional strategies to changing socio-economic interactions, *European Planning Studies*, 14(7), pp. 959–978.

Sijmons, D. (2002) *Landkaartmos en andere beschouwingen over het landschap* [Map Moss and Other Reflections about Landscape] (Rotterdam: 010 Publishers).

Sijmons, D. & Venema, H. (Eds) (1998) = *Landscape* (Amsterdam: Architectura & Natura).

Stalenberg, B. (2010) Design of flood-proof urban riverfronts, PhD thesis, Technische Universiteit Delft, Delft.

Stuurgroep Zuidwestelijke Delta (2009) *Vraag en aanbod van zoetwater in de Zuidwestelijke delta—een verkenning* [Supply and Demand of Fresh Water in the Southwestern Delta: An Exploration] (Utrecht: Kennis voor Klimaat).

Throgmorton, J. A. (1992) Planning as persuasive storytelling about the future: Negotiating an electric power rate settlement in Illinois, *Journal of Planning Education and Research*, 12, pp. 17–31.

Tjallingii, S. (1996) *Ecological Conditions; Strategies and Structures in Environmental Planning* (Wageningen: Wageningen University Press).

Van De Ven, G. P. (2004) *Man-made lowlands; History of water management and land reclamation in the Netherlands* (Utrecht: Matrijs).

Van Den Brink, G. (2007) *Prachtwijken? Mogelijkheden en beperkingen van Nederlandse probleemwijken*, [Strong Neighbourhoods? The Potential and Limitations of Dutch Urban Problem Areas] (Amsterdam: Bert Bakker).

Van Duinen, L. (2004) Planning imagery: the emergence and development of new planning concepts in Dutch national spatial policy, PhD, University of Amsterdam, Amsterdam.

WWF (World Wildlife Fund) (2008) *Hoogtij voor Laag Nederland: Werken met de natuur voor een veilige en mooie delta* [High Tide for Netherlands: Working with nature in the interest of a safe and beautiful delta] (Zeist: WWF).

WWF (World Wildlife Fund) (2010) *Met Open Armen. Voor het belang van natuur, veiligheid en economie* [With Open Arms. In the Interest of Nature, Safety and the Economy] (Zeist: WWF).

Improving Sustainability in Urban Areas: Discussing the Potential for Transforming Conventional Car-based Travel into Electric Mobility

BERT VAN WEE*, KEES MAAT** & CEES DE BONT†

*Policy and Management, Section Transport and Logistics, Delft University of Technology, The Netherlands,
**OTB Research Institute for the Built Environment, Delft University of Technology, The Netherlands,
†Faculty of Industrial Design Engineering, Delft University of Technology, The Netherlands

ABSTRACT *To reduce the environmental burden of urban road traffic, researchers and policy-makers have focused on reducing car use in urban areas, and improving public transport. Current expectations with respect to electric cars are high, not only for environmental reasons, but also because of the expected depletion of fossil fuels. Using a conceptual model for environmental burden caused by cars and a categorization of policy instruments and their effects, this paper discusses policies to encourage the adoption of electric cars. The paper also discusses the implications of the successful introduction of electric cars for conventional urban transport policies. The paper concludes that electric vehicles could bring substantial benefits to the environment and energy consumption, but the long-term environmental benefits largely depend on the size of potential rebound effects, the life-cycle effects with respect to energy, differences between electric cars and their competitors in "well-to-wheel" consumption, and—more particularly—improvements with respect to its main competitor: improved internal combustion engine powered cars. Secondly, the paper shows that policies to encourage electric cars should be adaptive. Probably in the early stages of introduction, privileges in central urban areas are useful. But in the long run, driving and parking electric vehicles should be limited for reasons of liveability. Thirdly, electric cars will never become a full substitute for public transport, the main reasons for this being land take and inefficient use of infrastructure.*

1. Introduction

Over the past four decades, the issue of environmental impact in general and the environmental impact of transport in particular has become increasingly important in policy-making and research. Over the past two decades, and particularly since about 2005, concerns related to climate change and the depletion of fossil fuels have grown. Transport is often viewed as the sector which makes the most significant impact on the environment (see Section 2). The transport sector depends heavily on fossil fuels. Both in order to reduce oil dependency and to minimize the impact on the environment, several technological options have been discussed in literature and policy-making over the past two decades. Beginning with policies announced in California with respect to Zero Emission Vehicles in the early 1990s, electric cars have been seen as a promising form of technology. However, due to a lack of progress in battery technology, these policies were never implemented and attention shifted towards hydrogen, bio fuels and very efficient internal combustion engine (ICE) technologies. However, in the past few years, electric vehicles (cars and other road vehicles) have returned to the policy and research agenda, particularly due to recent improvements in battery technologies. Car manufacturers and other automotive industries are investing heavily in R&D in this area. At the time of writing, the first electric cars are available for purchase and several large car manufacturers have announced that they will introduce electric cars onto the market within a few years.

From the perspective of climate change and sustainable cities, there are some important questions related to the future of electric cars:

(1) What are the environmental pros and cons of electric cars in general, and with respect to the urban environment in particular?
(2) What dynamics can be expected with respect to the introduction and diffusion of electric cars, and what role could governmental policy play in their introduction and diffusion?
(3) Which relationships exist between (policies related to) the introduction and diffusion of electric cars, and conventional policies to improve the urban environment, in particular public transport policies, and policies to reduce car use?

We aim to answer these questions from a planning perspective. As a result, our focus here is more on planning-related issues than on technological issues. We aim to answer these questions on the basis of literature and analytical reasoning.

Section 2 gives an overview of the relationship between transport and the environment. Section 3 presents a conceptual model for the environment effects of the transport system that are relevant to the policy. Section 4 discusses policy instruments and their potential effects on determinants for the impact of transport on the urban environment. Section 5 is the core of the paper. It discusses electric vehicles, using the conceptual model presented in Section 3 and the policy instruments presented in Section 4. Finally, Section 6 summarizes the main conclusions of this paper.

2. The Environmental Effects of Transport[1]

This section aims to give an impression of the impact of the transport sector on the environment, and to make clear why transport is often seen as the sector which has the greatest adverse effect on the environment. In brief, this is because the transport sector accounts for a high proportion of the emissions of several pollutants. Table 1 shows the share of total emissions of SO_2, NO_x, CO, HC and CO_2, that are accounted for by the transport sector in the US. As shown, in the US, the transport sector is the source of

Table 1. Share of transport in emissions in the US, 2008

SO_2	4.5
NO_x	57.9
CO	73.2
VOC	37.7
PM-2.5	0.2
PM-10	3.2
CO_2	33.2

Source: Davis *et al.* (2010).

the majority of NO_x and CO emissions, over one-third of HC emissions, and one-third of CO_2 emissions. The share of SO_2 appears low, but emissions from marine shipping are excluded from these figures, because these do not occur within a country's boundaries and are therefore not included in national statistics.

However, Table 1 does not capture the extent to which transport impacts on human health because, on average, the distance between road traffic and those exposed to the pollutants emitted is much shorter than for other sources of pollution such as power plants. Traffic emissions therefore have a greater health impact per kilogramme of pollutant emitted than average emissions (Dorland & Jansen, 1997; Eyre *et al.*, 1997; Newton, 1997). This applies particularly to road transport—more so than for non-road transport—and the effect is even stronger in urban environments. The importance of the distance between the source of the pollutant and the receptor is expressed by the concentration of the pollutant as a function of the distance between a given location and a given road: the greater the distance, the lower the concentration. Bennett *et al.* (2002) have introduced the concept of the "intake factor", which is the intake of a pollutant by individuals and expressed in mass, divided by the total mass of emissions. Evans *et al.* (2002) give an overview of studies on this subject and refer to a study by Smith (1993a, 1993b) which concludes that the intake of particulates emitted by vehicles is 10 times higher than those emitted by a power plant. For further examples of this concept, see Marshall *et al.* (2003, 2005). To conclude, the share in emissions from different sectors is only a rough indicator of that sector's share in the health effects of that pollution. The impact of distance on the effect of pollutants should be taken into account if we are to gain insight into the environmental effect of emissions, resulting in a relatively high share of road transport emissions.

Forecasts show that emissions of "traditional" pollutants (CO, HC, NO_x, PM) are expected to decrease, whereas an increase in CO_2 emissions is generally expected (WBCSD, 2004).

3. A Conceptual Model for the Effects of the Transport System on the Environment, Accessibility and Safety[2]

Before we can discuss the potential pros and cons of electric cars, it is important to understand the factors that contribute to the impact of transport on the environment. In this section, we describe a conceptual model for the effects of the transport system on the environment, accessibility and safety. This model is presented in Figure 1. We start by explaining trends in transport volumes. Given the population size and its break-down

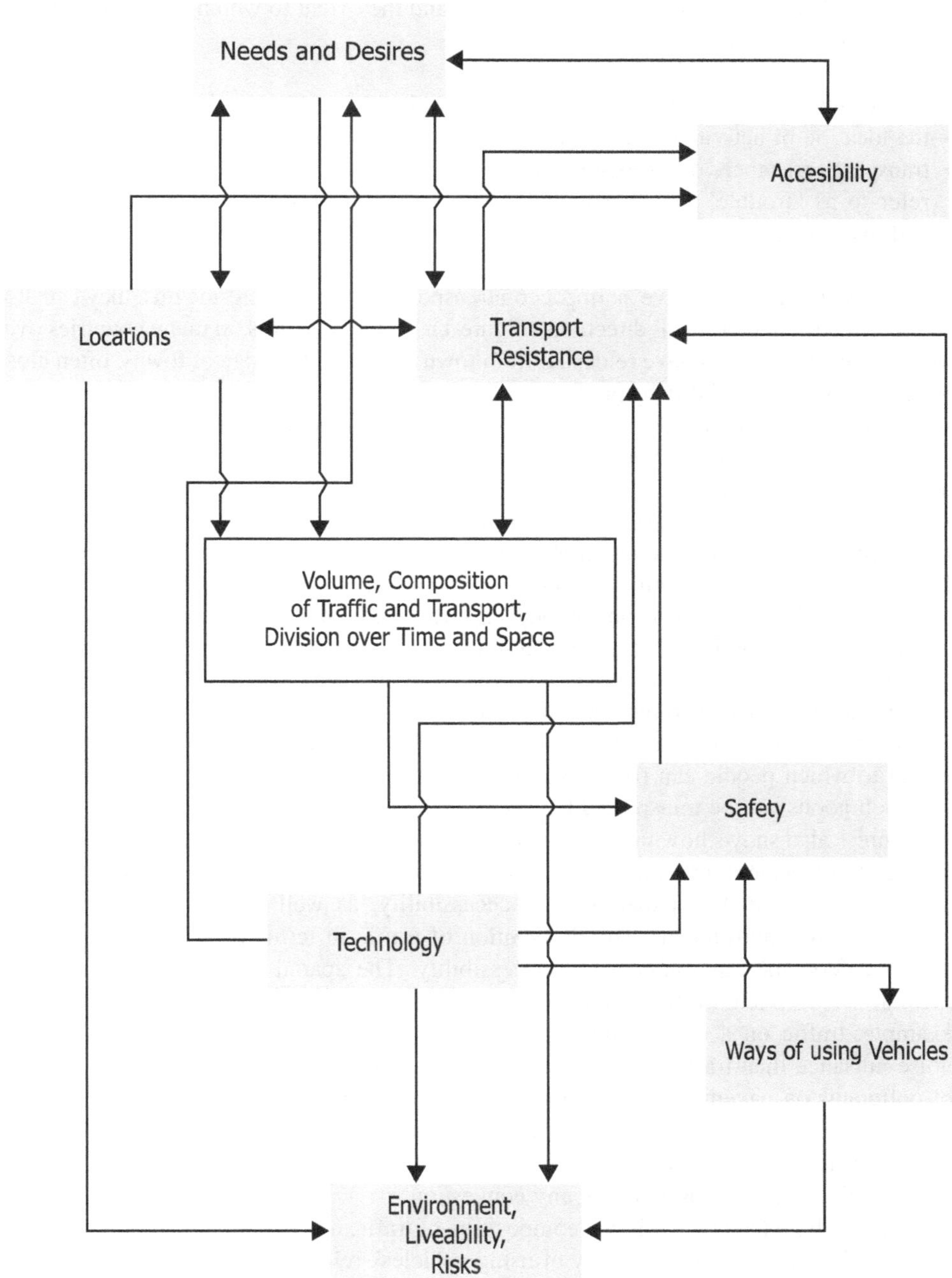

Figure 1. A conceptual framework for the effects of the transport system on the environment, accessibility and safety.
Source: van Wee (2009).

by household class and age, transport volumes and the extent to which the various modes and vehicle types are used are the result of:

- the desires, needs, preferences and options available to people and firms;
- the location of activities such as living, working and shopping;
- transport resistance, often expressed in time, financial cost and other factors, that we refer to as "trouble" and that include risks, the unreliability of the transport system and dis-effort.

All these three factors have an impact on transport volumes. Alongside this, they can also impact on one another, in all directions (Figure 1). To illustrate this, in many countries over the past decades, offices have relocated from town centres to the edge of towns, often close to motorways. Accessibility by public transport has thus decreased while accessibility by car has increased. In other words, changes in locations have an impact on transport resistance to travelling by car and public transport and may result in an increase in people's desire to own a car or a second car. Because all three categories of factors change continuously, a perfectly stable equilibrium between the three categories of factors is not possible. What is more, the relationships between the factors imply that a policy that focuses exclusively on one of the factors may have several indirect effects. For example, if electric cars had much lower CO_2 emissions per kilometre than conventional cars, this would not only proportionally decrease car-related CO_2 emissions, but it would also decrease fuel costs per kilometre, and so boost car use—the so-called rebound effect. This may cause people to switch from other modes to electric cars (an effect also included in the rebound effect).

The three factors also have an impact on accessibility. We define accessibility as the extent to which people can participate in activities at different locations (and the extent to which goods can be transported between locations).

Figure 1 also shows how technology and driving behaviour (as expressed by speed and acceleration/deceleration behaviour) have an impact on travel times and travel comfort (transport resistance) and therefore on accessibility, as well as having an impact on safety and the environment. The distribution of traffic in terms of space and time also affects safety, the environment and accessibility. The spatial distribution includes the breakdown between traffic inside and outside built-up areas and by road class. For example, traffic on a road along which hardly any houses are located will cause less noise nuisance than traffic on a road where many houses are located. The concentration of pollutants on pavements will be higher if the pavement is located near a (busy) road. The time of day breakdown is also very relevant, since the noise nuisance caused by nocturnal traffic is much greater than that of daytime traffic. On the other hand, in most regions, night traffic causes hardly any congestion.

Note that both the box "volume, composition of traffic and transport, division over time and space" as well as the box "way of using vehicles" relates to behaviour. The first box expresses travel behaviour in terms of numbers of trips, distances, mode choices, route choices, and departure (and arrival) time, the latter box expresses the way drivers use vehicles in terms of variables such as speed, acceleration and deceleration.

To summarize, the effects of the transport system on the environment, accessibility and safety are the result of the complex and dynamic interaction of various aspects of the transport system. These factors and how they are interrelated are visualized in Figure 1. Only the dominant relationships are included.

4. A Categorization of Policy Measures

To gain a general overview of the possible factors that determine the impact of transport on the environment, accessibility and safety, the next important question is: what can policy-makers do to make the transport system more sustainable? This section gives an overview of possible policy measures. The impact of transport depends on a number of determinants (Figure 1). The first is the overall volume of transport, expressed in terms of passenger kilometres (persons) or tonne kilometres (goods). The second category is the modal split (for passengers this is car-driver, car passenger, train, bus/tram/metro, walking, cycling, aircraft, ship), and the third is the technology used. Technology relates predominantly to environmental and safety characteristics. In the future, vehicle technology may contribute to improvements in accessibility by reducing congestion levels, particularly on motorways (e.g. the introduction of technologies to enable cars to drive closer together by shifting driving tasks from the driver to technology). A fourth category is the efficiency of using vehicles (for freight vehicles this is the load factor, and for cars, trains and buses it is the occupancy rate). The fifth is the manner in which vehicles are used (speed, acceleration and deceleration). Governments have several policy instruments through which they can seek to influence these determinants, including regulations, prices, land-use planning, infrastructure planning, marketing and the provision of information and communication. Table 2 shows the relationship between the type of policy instruments and the determinants.

Table 2 shows that regulations and pricing measures have the greatest potential impact, followed by land use and infrastructure measures. We will now give examples of the dominant effects as presented in Table 1, focusing on (urban) environmental impacts.

Restrictions can reduce the overall volume of traffic—by limiting access to certain areas, for example. Where restrictions are placed on access for certain types of vehicle (e.g. motorized transport, which is often the case in central urban areas, or electric vehicles), this can affect the modal split. Most technological changes in vehicles that reduce emissions or improve safety are the result of new regulations. Restrictions on large freight vehicles in certain areas may affect their load factors (and thus also the efficiency of using freight vehicles at all). The CAFE (Corporate Average Fuel Economy)

Table 2. Dominant relationships between transport policy instruments and determinants of sustainability

	Volume	Modal split	Technology	Efficiency of using vehicles	Use of vehicles/ driving behaviour
Restrictions	•	•	•	•	•
Pricing policies	•	•	•	•	•
Land-use planning	•	•			•
Public transport policies	•	•		•	
Infrastructure	•	•			•
Marketing		•			
Information and communication	•	•		•	•

Source: Based on Blok and van Wee (1994).

regulations in the US have reduced the energy use of cars. Speed limits also reduce energy use and some emissions, as well as improving safety.

Pricing policies can also have a range of impacts. For example, increasing levies on fuel may reduce overall transport volume, and lead to a shift to other modes. Higher taxes for vehicles without a three-way catalytic converter in the years before 1993 (the year in which the Euro 1 standards were introduced), increased the sales of "clean" cars in Germany and the Netherlands in those years and led to technological changes in the car fleet. Higher fuel prices lead to a rise in the share of car usage (efficiency) and to lower speeds on motorways in order to save fuel.

Land-use planning is particularly important at the local and regional levels. More compact land use and mixed land use may reduce the overall volume of traffic, and increase the share of slower modes of transport. It may also change the distribution of car kilometres over road classes and therefore affect driving behaviour.

Providing more and faster *infrastructure* will increase transport volumes and driving speeds. Investing more in public transport may increase the number of trips made by public transport. Investing in recharge infrastructure for electric cars could accelerate their market penetration.

Marketing public transport may increase its share; marketing electric cars could increase their market share.

The provision of *information and communication* may change people's travel behaviour. For example, route information may reduce trip distances, information about public transport may increase its share. Information provided through ICT may increase the efficiency of freight transport by making it possible to pick up a load on a return trip. Information provided by on-board units may change people's driving behaviour.

5. Electric Cars in Urban Areas: A Discussion

We will now turn to the questions raised in the introduction, using the conceptual model of Section 3, and Table 2. We will begin with the position of electric cars in the conceptual model. To understand the position of electric cars in Figure 1, Table 3 summarizes the main features to be discussed and gives an indication of the range of features. The range is based on (grey) literature, expert's opinions and information from car manufacturers and battery producers. Note that this range is based on the current (early 2011) state-of-the art. New technologies generally improve over time, and costs decrease, thanks to production on a larger scale ("scale effects") and learning ("learning effects").

5.1. *Understanding Behavioural Changes of Electric Cars Using the Conceptual Model of Figure 1*

If electric cars are introduced, this will directly affect transport resistance in a number of ways. The cost of electricity per kilometre will be significantly lower than fuel costs per kilometre, which—as expressed in Figure 1—may be an incentive to travel more. This will result firstly from the low electricity costs per kilometre. In addition, it is possible that electric cars will be used to generate and store energy for other purposes (via the "smart grid") allowing people to earn money if their cars (or, more specifically, the batteries) are used for these purposes. Secondly, (psychological) car costs per kilometre may decrease for particular groups, for example, people who prefer to minimize their

Table 3. Main features of electric cars compared to ICE cars

Feature	Indication for range
Lower costs per kilometre	Depending on levies on fuel: up to −95%
Fewer environmental disadvantages (in terms of CO_2 emissions, air quality and noise)	Depending on the way electricity is produced −30% to −100%
Lower range	Current range: 150–200 km. Then −50% to 75%. Future improvements in range can be expected
Long recharging times	Current technology: 15 minutes to 8 hours. Short recharging times may reduce lifetime of batteries
Changes in performance (faster acceleration, more weight, lower loading capacity)	Acceleration times: −25% to 60%. Weight increase: +25–75%. Loading capacity: −25% to 50%
Parking advantages in urban areas	Depending on specific advantages, e.g. spaces for electric cars only
Parking advantages at work location and other destinations	Depending on specific advantages, e.g. spaces for electric cars only, spaces for electric cars closer to the entrance of buildings

car use for environmental reasons may have less reason to continue to do so. This is because of relatively low CO_2 emission of electric cars in general, and even more so when solar and wind-energy are used to generate electricity, and also because these vehicles do not pollute locally and make hardly any noise when driven at speeds of up to 50 km/h (above that speed, tyre noise is greater than engine noise, as expressed in the regulations for acoustic research of many Western countries). This lower resistance could lead to an increase in overall travel, as well as a shift from public transport or slower modes to the electric car. Furthermore, many firms interested in reducing their carbon footprint might consider switching to buying and hiring more electrical vehicles (EVs) rather than using traditional modes to move people and goods.

On the other hand, electric cars have characteristics that could increase transport resistance, the effects of which would be the reverse of those explained above. First the range of EVs will probably be limited compared with ICE cars, unless there are breakthroughs in battery technology. Secondly, recharging takes much longer than refuelling a car. Again, technological progress could change this. On the other hand, if recharging at work becomes possible, for example, this would increase the convenience of recharging, although the recharging time will still be longer than refuelling a car. Third, it is possible that recharging will not be needed because (standardized) battery packs can be changed if empty, for example, at battery-swap stations. In the long term, it is likely that technological advances will increase the range of EVs. In the meantime, hybrid electric cars with a smaller battery pack than all-electric cars and a combustion engine as a range extender have also been introduced (e.g. Toyota Prius plug-in hybrid and Opel Ampera) to address range anxiety.

Resistance could further be affected by the performance of electric cars which will very likely differ from that of ICE cars. Electric cars accelerate faster than ICE cars because power is immediately available (and does not come through an increase in engine revolutions). On the other hand, the maximum speed of the cars currently available or announced in the coming years is lower than that of comparable ICE models. This is because at higher speeds, electric vehicles can travel much less far before recharging becomes necessary, and in some cases because high speeds may also lead to battery overheating. However, since electric cars can reach maximum speeds that are generally (significantly) above the maximum speed limit in most Western countries, we do not believe that the lower maximum speed of electric cars is a practically relevant issue, but possibly a psychological reason why some people may prefer an ICE car. Cars are probably an example of positional goods (Alpizar *et al.*, 2005), meaning that the utility of a car for its owner depends on characteristics of other cars in the fleet, the cars of neighbours, colleagues and friends being the anecdotal examples. Other characteristics of EVs also mean that they differ from ICE cars. If electric cars are based on ICE models, less luggage space would be available because of the storage of batteries. They are also heavier for the same reason, leading to different handling characteristics. Not all electric cars will be based on ICE models; some will be designed from scratch. It is possible that such cars will be relatively small to increase their energy efficiency and increasing their range. Such small cars will in general perform worse in terms of space and handling, but better in terms of range and manoeuvring (driving and parking) in urban areas. In addition, changes in resistance may occur if electric cars are given special privileges, such as allowing electric cars on roads where ICE cars are not allowed, special parking places in urban areas (either general parking, or parking with recharge facilities), special parking places at the work place (where they can be recharged) and other destinations, such as near parks or shopping facilities.

To summarize, resistance to the use of electric cars will change in a number of ways *directly*. Figure 1 shows that *indirect* relationships result from two relationships (represented by arrows): the arrows from “resistance” to “location”, and to “needs” and “desires”. The arrow to “locations” conceptualizes that people might change destinations. If they have electric cars, they may prefer locations or destinations within their range (if they can recharge at the destination) or at less than half of the range (if they cannot recharge). They may also prefer to do their shopping in the inner city because of parking privileges rather than on the outskirts of town. If travelling by electric car is less comfortable, they may also prefer to drive over shorter distances and change their destinations for that reason. Over the long term, changes may even occur in land use, as generally recognized in the literature on the interaction between land use and transport (Wegener & Fürst, 1999). For example, if accessibility changes, either positively or negatively, activity patterns will change, and this will have long-term consequences for the spatial structure. The other arrow is from “resistance” to “needs” and “desires”. It is possible that the availability of electric cars at a reasonable price may increase levels of car ownership (including an increase in multiple car households), due to an increase in the preference for transportation by car. Figure 1 also shows that accessibility, which is very relevant from the point of view of transport and land-use policies, may be affected by the introduction of electric cars.

5.2. *Policy Instruments*

Based on the categorization of Table 2, this section discusses policy instruments that could relate to enablers and barriers for electric cars.

In general, if governments would like to encourage electric mobility, their role is twofold: (1) it should ensure that EVs (with their disadvantages) become competitive with conventional cars, and (2) it should ensure that the necessary infrastructure is available, in particular charging facilities. To improve the position of EVs in the car market, the following policy instruments can be applied.

Restrictions can relate to the right to use certain roads and parking spaces, as mentioned above. To encourage the use of electric cars, they may be allowed to drive on certain roads or in certain areas, or park in spaces that ICE cars are not allowed to park in, and they can be provided with recharging facilities. Compared with the current situation, electric cars could be provided additional privileges, but it would also be possible to prevent ICE cars from accessing places where they are now allowed to access, either to stimulate electric cars, or to reduce air pollution or for other liveability reasons. Restrictions could also relate to the maximum speeds at which new cars can drive. If, for example, new cars were prevented from driving faster than the maximum speed on motorways, the difference in top speed between electric cars and ICE cars would disappear, improving the competitive position of electric cars.

Road pricing can be implemented at the national or urban level. At the national level, purchase and yearly taxes for electric and ICE cars could differ. On the other hand, the government would face a decrease in tax revenues if the use of EVs becomes widespread because the government would receive less tax revenues from fuel duties. To compensate for this, the central government could introduce a per-kilometre tax on electric cars. At the national level, pricing electricity is an option, certainly where electricity is produced and distributed by the public sector. To encourage recharging at night, electricity could be priced more cheaply at night. Policies could relate to encourage people to install solar panels in their homes to recharge cars. Even in countries with a private electricity sector, policies could be introduced with respect to buying and selling renewable energy as has been done in Germany (where over a long period, companies and people can sell electricity to private electricity companies for a guaranteed price).

Subsidies and tax advantages (temporarily, in the early stages of the market introduction, or permanent) are other options for pricing instruments to stimulate the purchase and use of electric cars. Tax advantages can relate to taxes on new cars, but also on yearly taxes.

Land-use planning a possible measure could be to select recharge locations for electric cars very carefully. A trade-off may exist between easy access for many users, and negative parking externalities. In large parking areas, such as at airports, the vast number of parked cars means there are opportunities to use car batteries as buffers for renewable energy sources, as described in Silvester *et al.* (2010).

Infrastructure policies could relate to parking and recharging. Theoretically a grid-wire infrastructure to provide cars with electricity is an option (Gilbert & Perl, 2007), but that would need heavy investment. Such investment in smart grids would contribute to wider developments in distributed energy generation in firms and households.

Marketing, information and communication can relate to electric cars and their use in general, and to its appearance in specific locations (such as on the Internet, in satellite navigation systems or personal digital assistant-based trip planners), and to provide people with up-to-date information on the availability of parking places with recharging facilities.

Combinations of policies could be attractive and lead to synergy effects. For example, when new recharge facilities are installed, marketing and information strategies may be needed to maximize their use. Combined land use, pricing, and restrictions policies could together make a strong case for the introduction of electric cars. One of the options for combining policies, dedicated to the stimulation of innovations as part of transition by public bodies, is strategic niche management (Kemp *et al.*, 1998). The basis of strategic niche management is the creation and/or management of niches for promising technologies. If strategic niche management were developed for electric cars, policies could, for example, include the purchase of EVs by governmental bodies, giving them parking privileges, and providing fiscal incentives.

It should be noted here that governments may fall into the well-known trap of lock-in, meaning that they encourage a sub-optimal solution, which will then beat the competition even though better alternatives exist. For example, EVs are now seen as a promising option, but if expectations surrounding battery improvements are not met, hydrogen cars or ICE cars that run on bio fuels may be more effective options. Government policies aimed at encouraging EVs should therefore not discourage other promising options at the same time.

5.3. *Success and Failure of Electric Cars*

Will electric cars become a success, reduce our dependency on oil, and improve the urban environment? This question is difficult to answer. Engineers emphasize the importance of breakthroughs in battery and recharging technology, economists focus on the purchase and per-kilometre prices of electric cars—both in absolute terms and compared with ICE cars, psychologists stress emotional factors and the importance of people's attitudes towards the environment and cars, sociologists point out the advantages of collective car ownership compared with individual car ownership in sharing the costs and risks involved in purchasing EVs, geographers and planners focus on the importance of parking and recharging facilities and where they are located. Most likely, all these specialists are right to some extent, and all those factors are also strongly interrelated. We have used the conceptual model of Feitelson and Salomon (2004) to conceptualize the factors that contribute to the success and failure of innovations in transport when the public sector plays a major role. The model was developed for innovations such as road pricing, light rail, and telecommunication centres, but is also very useful for electric vehicles. Figure 2 visualizes their model.

Feitelson and Salomon make it clear that the adoption of new technologies depends on technical feasibility and political feasibility. Those two categories of feasibility depend on many factors and in the case of electric cars, they include the opinions of experts, the interests of the industry (car manufacturers, the battery industry, electricity companies), the sense of urgency with respect to the availability of oil and climate change, costs and benefits (real and perceived), for both the users of electric cars, as well as for society. We will not discuss all the elements in the model. Instead, we will conclude that it is still rather uncertain whether electric cars will become a market success in the decades to come. In addition to technological developments related to electric cars, batteries, and recharging facilities, we need to understand the position of crucial actors (private and public) and their interests, the interaction between them, and the preferences and behaviour of consumers (and other potential users). Changes in technical and non-technical characteristics, the enablers and barriers for the success or failure of electric cars, will, most probably, interact.

Note that governments are not included as actors, because they are explicitly included in the box "political feasibility". The public is not explicitly included. Partly their impact runs via interest groups. However, the public may also play an independent role, for example, because they can vote, and social feasibility influences political feasibility. The impact of the public, in their role of voters and in other roles, and next the impact

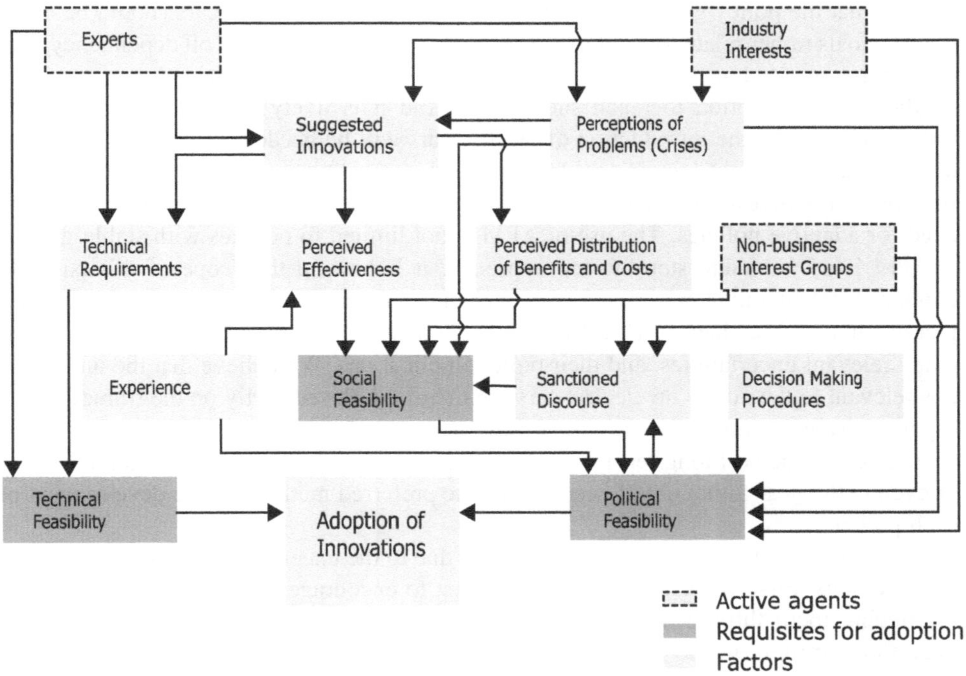

Figure 2. A political-economics conceptual model for the success or failure of innovations. *Source*: Feitelson and Salomon (2004).

on governments and politicians is explained in Public Choice Theory (Buchanan & Tullock, 1962).

5.4. *Dynamics and Adaptive Policy-Making*

Partly in relation to the complex system of factors that could contribute to the potential success or failure of electric cars, as conceptualized in Figure 2, the introduction and diffusion of electric cars will be a highly adaptive process and as such will demand adaptive policy-making (APM). APM is a concept developed for cases where the goals of policy are known, but in which there is large uncertainty about the future. Owing to that uncertainty, the policy needs to be adapted.

The concept is illustrated by a metaphor in Marchau *et al.* (2010):

> The basic concept of adaptation is easy to explain. It is analogous to the approach used in guiding a ship through a long ocean voyage. The goal—the endpoint—is set at the beginning of the journey. But, along the way, unpredictable storms and other traffic may interfere with the original trajectory. So, the policy—the specific route—is changed along the way. It is understood before the ship leaves port that some changes are likely to take place—and contingency plans may have already been formulated for some of the unpredictable events. The important thing is that the ultimate goal remains unchanged, and the policy actions implemented over time remain directed toward that goal. (If the goal is changed, a new plan must be developed.)

Note that the policy goal probably is not beforehand that electric cars should be introduced. Goals rather relate to a reduction of the use of fossil fuels (and oil dependency) and CO_2 emissions, and air pollution. On the other hand, electric cars can be one interesting candidate policy option to reach such goals. And it is likely that specific policies to reduce barriers for the introduction of electric cars will be needed for a successful introduction and diffusion. Changing goals, either with respect to energy, climate change or air pollution, or with respect to the desired position of electric cars will even more increase the need for adaptive policies. The use of APM is not limited to policies with stable goals.

APM is a systematic step-by-step approach. It is beyond the scope of this paper to explain APM in full. For a description and three illustrations in the area of transport, please refer to Marchau *et al.* (2010). What is particularly important is the notion of many relevant uncertainties, and their policy implications. We believe that the uncertainties relevant to the policy on electric cars are manifold. Based partly on the framework of Figure 2, Table 4 summarizes some of those uncertainties.

As a result, the best long-term policies relating to electric cars cannot be defined today or even in the near future. APM seems to be the preferred method for the development of such policies.

Dynamic policies may be preferred not only due to the uncertainties above, but also for other reasons. First, if (local) governments want to encourage the use of electric cars by privileging them through free parking, for example, such policies may be abandoned once those stimuli are no longer needed—maybe because the market share of electric cars increased rapidly. Such privileges could even be the victim of their own success: the numbers of electric cars may become too high to be able to continue privilege policies. Secondly, if one aspect of the policy is that governmental bodies will buy many electric cars in the early stages in order to encourage successful market penetration, such a policy may become redundant when the market starts booming. Third, in some cases, private parties may take over tasks from public parties, such as with respect to providing

Table 4. Principal uncertainties relevant to the policy on electric cars

Technological uncertainties	Breakthroughs in the area of technology (such as battery capacity/range, recharging, options for changing battery packs)
	Interaction with electricity production (including those related to developments in a smart grid)
	Breakthroughs in competing technologies such as hydrogen, bio fuels and very efficient ICE cars
Uncertainties in external factors	Future oil prices, depletion of oil reserves, stability of oil production
	Climate policies
Societal uncertainties	Safety issues (in high-voltage charging and car accidents)
	Acceptance by consumers
Economic uncertainties	Costs and prices of battery packs and electric vehicles (including residual value)
	Behaviour of current and potentially new car manufacturers
Policy uncertainties	Policies of other countries and the EU
	Local policies
	Policies of employers
	Recharging infrastructure (which depends on public policies and companies such as oil companies)

recharging infrastructure. Fourth, dynamic policies may be required for reasons of land use and liveability. If electric cars would have a very high market share, such as over 50% of the car fleet, substantial gains with respect to energy use and CO_2 emissions are possible. But problems related to land take (for driving and parked vehicles) and liveability issues related to the dominance of (electric) cars in the urban environment remain. In fact, these issues could even become more salient because of the privileges allowed for electric cars to access or park in areas where ICE cars had not been allowed. In that case, policies to address land take and liveability issues would still be needed, possibly even more urgently.

To conclude, due to both the significant uncertainties and the non-uncertainty related dynamics in the market position of electric cars, policies will need to be changeable over time.

5.5. *Implications for Traditional Policy Instruments*

The negative aspects of car use have prompted a series of measures to reduce car use. As a result of these measures, European countries are much less car dependent than, for example, the US. However, the private car has remained very attractive, and activity patterns have become partly dependent on the car as well as the urban structure. As a consequence, alternative modes, such as the bicycle and public transport will never become as important as the car. The most likely option to replace the conventional car is, then, another type of car, but a much more environmentally sustainable version.

Nevertheless, an important question remains: will current policies to reduce the negative effects of car use become obsolete if electric cars take a higher market share? To answer this question, we have to distinguish the reasons why such policies are implemented. In terms of CO_2 emissions, localized air pollution and energy policies, policies to promote electric cars could be much more effective than those to encourage the use of public transport or reduce car use (parking policies, limited access of cars), especially when in "well-to-wheel" terms, electric cars consume much less fossil fuels and emit much less CO_2 (e.g. because they use electricity produced by renewables). Here we wish to stress the importance of comparisons between modes in well-to-wheel terms. Such comparisons are often made on a per-kilometre basis only, but the production of fuels (from the "well", i.e. an oilfield, to the fuel station) requires energy itself and also causes emissions. The per-kilometre performance of electric cars (along with electric trains and trams, and metro) depends heavily on how the electricity is produced. On the other hand, if the availability of fossil fuels becomes really problematic, or if ambitious reduction targets for CO_2 are set, maybe all reasonable policy options will need to be introduced to meet the CO_2 targets, or to make transport systems function, using less oil.

Another reason for current policies is to reduce the localized negative effects of car use (mainly in urban areas), such as air pollution, noise, parking pressure, safety and reduced liveability due to heavily congested streets. Current policies to reduce these effects include the provision of public transport, subsidies for public transport, reducing access to cars, and parking policies. In some countries, including the Netherlands and Denmark, and increasingly in other countries such as Germany and France, cycling is being encouraged by the provision of cycle infrastructure. Electric cars have the potential to reduce local air pollution significantly because they do not emit pollutants at the local level (but via the production of electricity, so at a larger distance from concentrations of population). Local noise pollution can also be reduced because the engines produce no noise—at low speeds (<50 km/h), the engine noise is the main source of car noise production. But safety problems and reduced liveability will not be solved by electric cars. Public transport and cycling require much less land, and are therefore much more efficient means of transporting people within, to, and from central urban areas. We, therefore, conclude that encouraging the use of electric vehicles is unlikely to make existing urban policies to reduce the negative effects of car use obsolete.

6. Conclusions

We conclude that electric vehicles have the potential to bring substantial benefits for the environment and energy consumption. Moreover, it is much easier to change society towards electric mobility than to alternative modes. However, both the short-term and the long-term environmental benefits depend largely on the size of potential rebound effects, the life-cycle effects with respect to energy, the "well-to-wheel" impact, and the improvements with respect to the main competitor: improved ICE powered cars. Secondly, this paper has shown that policies to encourage electric cars should be adaptive. In the early stages of introduction, privileges in central urban areas will probably be useful. In the long run, driving and parking electric vehicles should be limited for reasons of liveability. Third, electric cars will not become a full substitute for public transport, the main reasons for this being the amount of land required and the inefficient use of infrastructure.

Notes

1. This section is partly based on van Wee (2011).
2. Sections 3 and 4 are based on several earlier publications of the first author, including van Wee (2009, 2011).

References

Alpizar, F., Carlsson, F. & Johansson-Stenman, O. (2005) How much do we care about absolute versus relative income and consumption? *Journal of Economic Behavior and Organization*, 56(3), pp. 405–421.

Bennett, D. H., Margni, M. D., Mckone, T. E. & Jolliet, O. (2002) Intake fraction for multimedia pollutants: A tool for life cycle analysis and comparative risk assessment, *Risk Analysis*, 22(5), pp. 905–918.

Blok, P. & van Wee, B. (1994) Het Verkeersvraagstuk, in: F. Dietz, W. Hafkamp & J. Van der Straaten (Eds) *Basisboek milieu-economie*, pp. 2215–2234 (Amsterdam: Boom) (in Dutch).

Buchanan, J. M. & Tullock, G. (1962) *The Calculus of Consent* (Ann Arbor, MI: University of Michigan Press).

Davis, S. C., Diegel, S. W. & Boundy, R. G. (2010) *Transportation Energy Data Book* (Oak Ridge, TN: Oak Ridge National Laboratory).

Dorland, C. & Jansen, H. M. A. (1997) *Extern Transport—The Netherlands. Dutch Case Studies on Transport Externalities* (Amsterdam: Institute for Environmental Studies (IVM), Free University).

Evans, J. S., Wolff, S. K., Phonboon, K., Levy, J. I. & Smith, K. R. (2002) Exposure efficiency: An idea whose time has come, *Chemosphere*, 49(9), pp. 1075–1091.

Eyre, N. J., Ozdemiroglu, E., Pearce, D. W. & Steele, P. (1997) Fuel and location effects on the damage costs of transport emissions, *Journal of Transport Economics and Policy*, 31(1), pp. 5–24.

Feitelson, E. & Salomon, I. (2004) The political economy of transport innovations, in: M. Beuthe, V. Himanen, A. Reggiani & L. Zamparini (Eds) *Transport Developments and Innovations in an Evolving World*, pp. 235–251 (Berlin: Springer).

Gilbert, R. & Perl, A. (2007) *Transport Revolutions. Moving People and Freight without Oil* (London: Earthscan).

Kemp, R., Schot, J. & Hoogma, R. (1998) Regime shifts to sustainability through processes of niche formation: The approach of strategic niche management, *Technology Analysis & Strategic Management*, 10(2), pp. 175–198.

Marchau, V. A. W. J., Walker, W. E. & van Wee, G. P. (2010) Dynamic adaptive transport policies for handling deep uncertainty, *Technological Forecasting and Social Change*, 77(6), pp. 940–950.

Marshall, J. D., Riley, W. J., Mckone, T. E. & Nazaroff, W. W. (2003) Intake fraction of primary pollutants: Motor vehicle emissions in the South Coast Air Basin, *Atmospheric Environment*, 37(24), pp. 3455–3468.

Marshall, J. D., McKone, T. E., Deaking, E. & Nazaroff, W. W. (2005) Inhalation of motor vehicle emissions: Effects of urban population and land area, *Atmospheric Environment*, 39(7), pp. 283–295.

Newton, P. N. (Ed.) (1997) *Reshaping Cities for a More Sustainable Future—Exploring the Link Between Urban Form, Air Quality, Energy and Greenhouse Gas Emissions*. Research Monograph 6 (Melbourne: Australian Housing and Research Institute (AHURI)).

Silvester, S., Beella, S. K., Van Timmeren, A., Bauer, P., Quist, J. N. & Van Dijk, S. J. (2010) *Integration of Electric Mobility into the Built Environment. Schiphol the Grounds*. Research Report, Delft University of Technology.

Smith, K. R. (1993a) Fuel combustion, air pollution, and health: The situation in developing countries, *Annual Review of Energy and the Environment*, 18, pp. 529–566.

Smith, K. R. (1993b) Taking the true measure of air pollution, *EPA Journal*, 19(4), pp. 6–8.

van Wee, B. (2009) Verkeer en vervoer: een introduction, in: B. van Wee & J. Anne Annema (Eds) *Verkeer en vervoer in hoofdlijnen*, pp. 15–28 (Bussum: Coutinho) (in Dutch).

van Wee, B. (2011) Urban transport and sustainability, in: E. van Bueren, H. van Bohemen, L. Itard & H. Visscher (Eds) *Sustainable Urban Environments: An Ecosystem Approach*, pp. 243–261 (Dordrecht: Spinger).

Wegener, M. & Fürst, F. (1999) *Land-use Transport Interaction: State of the Art. Deliverable D2a of the Project TRANSLAND (Integration of Transport and Land use Planning)*. (Dortmund: Berichte aus den Insititut für Raumplanung 46, Universität Dortmund, Insititut für Raumplanung).

World Business Council for Sustainable Development (2004) *Mobility 2030: Meeting the Challenges to Sustainability*. Available at http://www.wbcsd.org/web/publications/mobility/mobility-full.pdf (accessed 6 December 2011).

Towards Energy-neutral New Housing Developments. Municipal Climate Governance in The Netherlands

MILLY TAMBACH & HENK VISSCHER

OTB Research Institute for the Built Environment, Delft University of Technology, PO Box 5030, NL-2600 GA, The Netherlands

ABSTRACT *The Dutch government intends to sharpen the current energy performance standard of the national building regulations for new buildings (Energy Performance Coefficient) step by step to an energy-neutral level by 2020. In their climate policies, municipalities are preparing to meet this level for new buildings—some of them even earlier. Nevertheless, possibilities to impose energy performance levels, higher than formulated at national level, are limited for municipalities. This paper investigates, how and through which policy instruments Dutch municipal authorities can promote energy-neutral new housing developments. This research suggests the following municipal climate governance approach and instruments: (1) an integrated design strategy for sustainable energy systems throughout the city: in both urban areas and buildings; (2) market-party-selection in tendering procedures based on sustainability criteria, including the energy performance of dwellings; (3) encouragement of community engagement for the deployment of renewable energy sources; (4) (inter)regional cooperation among cities in similar climatic zones on the development of both more uniform calculation methods for zero-energy buildings, and of a more uniform methodology for cities to monitor progress towards the self-set targets. However, stable and long-term-oriented European and national financial support for municipalities seems to be needed for municipal climate governance to become successful.*

1. Introduction

In 2008, the Dutch housing stock consisted of around seven million dwellings (and almost 7.2 million dwellings in 2010: see CBS, 2011). For the coming years, the Dutch government aims to achieve an annual production of around 70,000 new dwellings, most of which will be located within or at the borders of cities and in middle-sized towns that will be experiencing considerable growth (cf. Todaro *et al.*, 2008a, 2008b; Tambach & Visscher, 2010). Most new housing estates will consist of combinations of social rented housing owned by non-profit housing associations, a major share of owner-occupied housing and a minor share of commercial rented housing. For example, new dwellings can be introduced and also reduce the environmental burden of an industrial estate within the city (Korthals Altes & Tambach, 2008). Although the energy performance of newly built dwellings has improved, the total average energy consumption of dwellings built after 1 January 2006 under the Dutch "Energy Performance Coefficient" (EPC) of 0.8 (sharpened to 0.6 as of 1 January 2011) and equipped with high-performance gas boilers, is still 20 GJ (Van de Griendt, 2010 and personal communication on 10 March 2010).

Given these developments, the energy performances of new dwellings and areas in which new developments are planned, will be of increasing importance, and adaptation to the consequences of climate change will have its implications for the planning, design and construction of new houses. In the debate on climate change, "mitigation approaches have been complemented by a new paradigm, that of adaptation to the risks introduced by climate change" (Kern & Alber, 2009). Lim *et al.* (2005) define adaptation as "a process by which individuals, communities and countries seek to cope with the consequences of climate change", whereas Adger (2003) describes the same as "a dynamic social process", as "the ability of societies to adapt is determined, in part, by the ability to act collectively". Within the field of economics, social capital is increasingly understood to have public and private elements, both of which are based on trust, reputation and reciprocal action (Adger, 2003).

Research on municipal climate governance, focusing on the planning of energy-neutral new housing developments has not received much attention in the international literature. One reason is, because the European policy goal of "nearly zero-energy building" established in Directive 2010/31/EU is relatively new, and countries need time to translate this goal into standards and national building codes. A second reason involves the existence of several barriers faced by municipal authorities in their efforts to achieve this goal. These barriers will be further elaborated in this introduction. As a third reason, many options for climate reduction are not used, even when municipalities have the competence to use them (Burger *et al.*, 2001; Coenen & Menkveld, 2002; Kern & Alber, 2009). Finally, local adaptation plans have only recently begun to be developed, and research on local adaptation strategies is limited (Kern & Alber, 2009).

Directive 2010/31/EU (EC, 2010) establishes "nearly zero-energy building" as policy goal for all new buildings by 2020. This directive was adopted by the European Parliament and the Council of the European Union to strengthen the energy performance requirements and to clarify and streamline some of the provisions from Directive 2002/91/EC (EC, 2003) it replaces. Directive 2002/91/EC (EC, 2003), also known as Energy Performance of Buildings Directive (EPBD), was intended to promote the improvement of the energy performance of buildings within the European Community. The new EU directive (2010/31/EU) states that Member States should draw up national plans for increasing the number of *nearly zero-energy buildings* (ZEBs) with very high energy performance, which require "nearly zero or a very low amount of energy that should be covered to a very significant extent by energy from renewable sources, including energy from renewable sources produced on-site or nearby" (Article 12, 2.).

In line with the new EU directive (2010/31/EU), the Dutch government intends to increase the energy performance standard for new buildings (EPC) step by step to an energy-neutral level by 2020 (MVROM, 2007). Whereas minimum physical requirements are determined at the national level, municipalities have a major influence on planning aspects and the architectural appearance of dwellings. Today, Dutch municipalities are in the process of formulating climate policies, which may imply that building regulations based upon the Dutch Building Decree (2003),[1] in particular the current EPC for new dwellings, need to be surpassed.

Local authorities are faced with barriers that prevent them to structurally implement options for GHG reduction in their daily practice. These barriers can be political (e.g. a lack of political support for a particular policy option), communicative (e.g. policy directions or ambitions being unclear), organizational or institutional (e.g. a lack of co-operation between different departments in the same local authority or dependence of local authorities on co-operation with other actors) or based on resources (e.g. human resources, finances, and low expectations of municipalities for certain options) (Burger *et al.*, 2001). The existence of institutional barriers has been highlighted by Meeus *et al.* (2011), Bulkeley *et al.* (2009), Davoudi *et al.* (2009), Bai (2007) and Tambach *et al.* (2010). Economic and information barriers have been emphasized by Meeus *et al.* (2011) and Tambach *et al.* (2010). Moreover, findings by Kern and Alber (2009) suggest that, in most countries, there is no direct link between national goals for GHG reduction derived from national climate protection programmes and international agreements (UN, EU), and the implementation of these goals at the regional and the local level (Kern & Alber, 2009).

In the Netherlands, an additional legal barrier is formed by Article 122 of the Dutch Housing Act (1901) revised in 1991 and 2007, which prohibits municipalities from undertaking acts of civil law regarding issues covered by the current Dutch Building Decree (2003), and this includes the EPC for new dwellings (cf. Hoekstra & Van der Veen, 2008; Tambach & Visscher, 2010). The aforementioned article also prohibits municipalities from sharpening Building Decree standards, such as the EPC, in local acts ("gemeentelijke verordeningen") (cf. Fieten, 2008). Although Dutch municipal authorities can reach voluntary agreements with developers on a higher EPC for new dwellings, this does not alter the fact that outside the recently appointed "excellent areas" (see Section 3) they cannot impose an EPC, surpassing the current standard (Tambach & Visscher, 2010).

The ZEB concept is regarded "a realistic solution for the mitigation of CO_2 emissions and/or the reduction of energy use in the building sector" (Marszal *et al.*, 2011) in an increasing number of countries, as witnessed by demonstration projects. One example is the Austrian "Active House" concept, in which the total energy demand of a building is supplied by renewable sources that are integrated in the building or from the nearby collective energy system and the electricity grid. Research on the Danish, Finnish, Norwegian and Swedish understanding of ZEB by Marszal *et al.* (2010) indicates that Denmark and Norway have established research centres that focus on ZEB concepts, whereas in Sweden and Finland, ZEB development is in an early stage.

According to Alsema *et al.* (2009), "a project is 'energy-neutral' if on an annual basis, no net import of fossil or nuclear fuels outside the system boundary is necessary to erect, use and demolish a building", and "a building (or an assembly of buildings) is 'CO_2 or climate neutral' if on an annual basis, no net GHG emissions are necessary to erect, use and demolish a building". Aspects relevant to this definition are the *project boundary* (all buildings and matching installations for energy conversion or CO_2-capture and -storage inside the direct sphere of influence of the project owner), the *system boundary* (the project plus various installations for energy conversion or CO_2-capture and -storage outside the direct sphere of influence of the project owner, but necessary for operating the buildings) and the technical and organizational measures taken for energy saving or—generation or CO_2-compensation inside and outside the project (but within system) boundaries (Alsema *et al.*, 2009). Despite the existence of voluntary environmental-assessment methods and labels such as GPR©gebouw and BREEAM, a clear and consistent definition and a commonly agreed energy calculation methodology of the ZEB concept are still missing (Marszal *et al.*, 2011).

This research was conducted within the framework of the EU URBACT II project HOPUS (Housing Praxis for Urban Sustainability), focusing on design and sustainability in housing and its regulation through municipal guidance and control. The research question in this paper is: *How and through which policy instruments can Dutch municipal authorities promote energy-neutral new housing developments*? Because of a missing clear and consistent definition and a commonly agreed energy calculation methodology of ZEB (Marszal *et al.*, 2011), and missing municipal instruments to develop such buildings on a large scale in European cities, the aim of our research is to analyze municipal climate-governance approaches and instruments that can also be used by city governments in other countries. Besides, the Netherlands is regarded as one of Europe's "leading" countries in the area of environmental policy (OECD, 1995; Janicke & Weidner, 1997). The research comprised a literature review, a document study and case studies, including interviews with civil servants in municipal housing and development departments, developers (housing associations) and an environmental manager in a development firm.

Section 2 describes the role of local authorities in climate policy and presents developments in municipal climate governance within an international perspective. Section 3 presents the Dutch government's energy policy for new dwellings, important agreements on the energy performance of new dwellings and related financial support schemes. In addition, it presents international climate-support programmes. Section 4 describes the municipal climate policies and instruments that are being applied for the development of highly energy-efficient new houses in locations in Amsterdam and Almere. Section 5 discusses the findings of both case studies as well as the national and European financial support provided to municipalities. The most important conclusions and recommendations are presented in Section 6.

2. Role of Local Authorities and Municipal Climate Governance: An International Perspective

2.1. *Role of Local Authorities in Climate Policy*

Local authorities are the tier of government that is closest to the citizen. They can play a major role in supporting and executing national climate policy, are responsible for local activities, resulting in GHG emissions and have responsibility in many policy fields that are significant for climate policy, such as town and country planning, transport, waste and energy management and have to issue permits for any new construction or form of land use (cf. Coenen & Menkveld, 2002). Most municipalities have the authority to include climate policy goals in their strategic planning, but are not obliged to do so—climate change policy remains for the most part a voluntary task of local authorities (Kern & Alber, 2009). The majority of Dutch municipalities have an energy policy since the 1970s and in the beginning, the main aim of this policy was to conserve energy, but little attention was paid to the role, municipalities could play with respect to other energy users like citizens (Coenen & Menkveld, 2002). Starting from the end of the 1970s, the way, municipalities performed their environmental tasks was discussed by both national and municipal government and apart from a general lack of priority, given to local environmental policy and a lack of local capacity in terms of manpower, major problems were found in a lack of local financial resources. During the 1990s, municipalities received earmarked funding to bring their environmental policy to a basic and adequate level (Coenen & Menkveld, 2002). In the first National Environmental Policy Plan (NEPP) (1989) and in the first white paper on energy saving (1990), municipalities were given a role as co-executor of energy policy, linked to sustainable development (Coenen & Menkveld, 2002).

Climate change is an example of multi-level decision-making (Collier, 1997; Coenen & Menkveld, 2002; Betsill & Bulkeley, 2006). Many cities around the world have developed their own climate action plans and strategies, but while mitigation was the main focus in the early phase of local climate change policy, both mitigation and adaptation measures can now be found on the local political agenda (Kern & Alber, 2009). Global climate change affects local governments in different ways, for example, through the high and increasing levels of GHG emissions generated in cities, and its impact at the local level varies among cities as well as the perception of these regional vulnerabilities and risks by citizens and policy-makers (Kern & Alber, 2009). In collaboration with its partners, the "Climate Alliance" (the European network of local authorities, committed to the protection of the world's climate) conducted a survey of municipal activities aimed at reducing CO_2 emissions in Austria, Germany, Italy and the Netherlands (cf. Coenen & Menkveld, 2002). The report concluded that the influence of the national framework is highly relevant to the extent and intensity of municipal climate-protection activities and that, in some cases, there were severe legal obstacles to further activities, while in most cases, the competencies of municipalities are not clearly defined or there is a lack of funding or know-how (Coenen & Menkveld, 2002). The playing field, i.e. the framework in which local authorities have to operate, differs per country and can alter in the course of time: the substance of local climate policy depends on the specific nature of the community in question, such as its geography, demography, economy, society and culture (Lafferty & Eckerberg, 1998; Coenen & Menkveld, 2002).

The framework employed for analysing and comparing modes of (urban) climate governance applied by the municipalities and in the cases studied in this paper (Section 4) is formed by the four governance modes introduced by Bulkeley and Kern (2006) and elaborated by Kern and Alber (2009): (1) *self-governing* is the capacity of the local government to govern its own activities, for example, by improving energy efficiency in municipality-owned real estate. It relies on reorganization, institutional innovation and strategic investment, (2) *governing through enabling,* referring to the role of local government in co-ordinating and facilitating partnerships with private entities and encouraging community engagement, largely by persuasion and (positive) incentives, (3) *governing by provision,* which implies shaping practice through the delivery of particular forms of services and resources, accomplished through infrastructure and financial policy, (4) *governing by regulation,* characterized as the use of traditional forms of authority, such as control and the use of sanctions. These modes of municipal (or urban) climate governance may overlap and policy instruments are often based on a combination of several modes.

In the field of mitigation policy, new approaches to energy policy have become the most visible instruments in local climate policy and aim primarily to improve energy efficiency and promote renewable energy solutions (Kern & Alber, 2009). Urban planning and land use provide local authorities with opportunities for local action, ranging from standards for new buildings to strategic planning for new neighbourhoods (Kern & Alber, 2009; see also Davoudi *et al.*, 2009; Schroeder & Bulkeley, 2009). In their Table 1, titled "Modes of urban climate governance and sectors of climate change mitigation", Kern and Alber (2009) present the mentioned four governance modes with coherent tools for four sectors (Energy, Transport, Waste, and Urban planning and land use).

The following governance-through-enabling-tools in the classification of Kern and Alber (2009, Table 1), play an important role in the analyses of the casestudy results in Section 4: "Campaigns for energy efficiency, Advice on energy efficiency to businesses and citizens, Promotion of the use of renewable energy, and Guidance for architects and developers on energy efficiency and renewables" (Kern and Alber, 2009, Table 1).

In many northern European countries, municipalities have specific competencies for the direct *provision* of energy, waste and transport services. But although they can have a significant capacity to address climate change that other local authorities lack, many of them seem to go beyond their delegated competencies in taking on responsibilities for providing services to their residents and businesses (Bulkeley & Kern, 2006; Bai, 2007; Granberg & Elander, 2007; Bulkeley *et al.*, 2009). Forms of horizontal and vertical collaboration are relevant to the success of urban climate governance beyond the local level (cf. Kern & Alber, 2009): *horizontal collaboration* within metro regions and city networking by knowledge and best-practice-exchange and a joint search for innovative solutions though often limited to innovators. But there is also *vertical collaboration* within nation-states, including various modes of governing, ranging from an *enabling* (by guidelines, awards, benchmarking and certification schemes) to a *providing* or *authoritative mode of governing* by the national government (Kern & Alber, 2009).

3. Dutch National and International Policies and Plans for Energy-neutral New Construction and the Available Financial Support

3.1. *Dutch National Policies, Plans and Financial Support Programmes*

In the Netherlands, energy makes part of sustainable building policy, as formulated in the policy letter sustainable building of 16 April 2002, which covers the following three spearheads: energy, materials and health. The Dutch cabinet's approach to sustainable development aims to make buildings and urban development sustainable by adopting innovative new-construction processes, stating that all new developments must be energy neutral by 2020 (MVROM, 2008).

The public-private Energy Transition Plan PeGO (PeGO, 2007) aims to achieve an 80% reduction of fossil fuel consumption in all buildings by 2050 (compared with 1990 levels) (cf. Tambach *et al.*, 2010). A first UKR (Unieke Kansen Regeling—Regulation for Unique Chances) subsidy tender "Towards energy-neutral housing" (2008) granted up to €0.5 million to municipalities, project developers and housing associations for new and innovative housing (or renovation) projects, comprising 50 dwellings with an EPC of 0.55 to the maximum, and suitable for scaling up. The CO_2-reduction of the project should be at least 45%, compared with the standard energy consumption in 1990 (SenterNovem, 2008). The tender was open for proposals for several months, and in the end there were 15 winning housing projects.

The Dutch government intends to increase the energy performance standard for new dwellings (home construction) step by step to an energy-neutral level by 2020 (MVROM, 2007). To reach this aim, their EPC of 0.8 was sharpened by 25% per January 1st 2011 and will be sharpened by 50% in 2015. In addition, the Dutch government has signed the following two agreements,[2] which affect municipal authorities:

(1) First, the 2007 *Climate Agreement* (VNG *et al.*, 2007) with the Association of Dutch Municipalities (VNG), in which signatory parties agree to promote that both national government and municipalities realize a substantial energy saving. VNG subscribes to the aim of energy-neutral new construction and a reduction of dwellings' and other buildings' energy consumption with more than 50% by 2020. Signatories agreed to give a good example in this themselves, and the step-by-step sharpening of the EPC for new dwellings is emphasized (Article 10, paragraph 1).
(2) Second, the 2008 *Spring Agreement* with branch organizations of the construction industry (Bouwend Nederland), project developers (NEPROM) and developers and building contractors (NVB), aiming at a reduction of the standardized energy consumption in the entire new building production[3] of 25% by 2011 and 50% by 2015, and at the development of a new energy performance norm. The agreement speaks of 10 areas to be selected for experimenting with new developments, at least surpassing the agreement's ambitions by 25%, on a large scale, and including 5–10% of the total annual new building production.

An example of the national government as "*provider*" in the sense of Kern and Alber (2009) is found in the provision of the SLoK (Stimulering Lokale Klimaatinitiatieven—Stimulation of Local Climate Initiatives) subsidy. This subsidy scheme, which succeeded the BANS (Bestuursakkoord Nieuwe Stijl—New Style Management Agreement) climate subsidy scheme,[4] largely supports the targets agreed upon in the *Climate Agreement.* Running from 2008 until 2013, it was introduced to stimulate municipalities and provinces to undertake structural activities and measures aimed at reducing GHG emissions (Tambach *et al.*, 2010).

The SLoK subsidy intends to reduce project costs for personnel, research, communication and education. Municipal governments can apply for subsidy to execute municipal projects, which meet certain programme performance targets of the Dutch "performance card for local climate policy" (1) in GHG reduction, energy saving and preservation, (2) per theme and (3) given certain preconditions to strengthen organizations—ambition levels ranging from "active" to "forerunning" and "innovative". With regard to the theme "energy efficient built environment—dwellings" municipalities can choose for an innovative ambition level, implying they need to realize either dwellings with a 75% sharpened EPC or energy-neutral dwellings and an EPL of 9 to 10, if housing projects entail more than 200 dwellings. A requirement to obtain the SLoK subsidy is municipal co-financing of at least 50% (Tambach *et al.*, 2010). Initially, €31.5 million was reserved for municipal authorities, but an additional €10.8 million needed to be injected by the Ministry of Housing, Spatial Planning and the Environment (VROM) (Schepers, 2009).

Stemming from the *Spring Agreement*, municipal governments were able to apply for grants up to €0.3 million per project for extra process and support costs (Van der Laan, 2009a). 13 projects (the maximum amount for this grant) were finally appointed as "excellent areas" by the Minister of Housing, Communities and Integration, in which municipalities and market parties can experiment in building highly energy-efficient and innovative new houses. A project needed to contain new dwellings with an EPC, surpassing the current standard: for example by at least 25%, if building works started before 1 January 2011. Municipalities must work together with developing parties, and in the excellent areas, they are allowed to enforce a higher than current EPC, upon which they have agreed with developing parties. The way in which compliance and enforcement of a higher EPC standard is regulated, is therefore a further selection criterion. In addition, projects needed to entail at least 1500 new dwellings (Van der Laan, 2009b), and have a learning effect for other areas. Innovations were possible in the field of technology, organization or applied (for example economic) instruments.

3.2. *International Climate-support Programmes and Initiatives*

In Sweden, local climate policies have been supported by LIP, a general environmental subsidy programme aimed at increasing ecological sustainability in society and creating new jobs started in 1998, and by its successor KLIMP (since 2003), a climate investment programme for climate projects aimed at the reduction of GHG emissions) (cf. Kern & Alber, 2009). Municipal co-funding is a requirement of both programmes. Kern and Alber (2009) state that between 1998 and 2008, approximately SEK (Swedish krona) 25 billion, of which about SEK 6 billion provided by the government, was spent on environmental and climate protection projects, and 126 climate- and 211 local- investment-programmes (2700 projects), have been funded. In 2008, the Swedish Environmental Protection Agency evaluated both programmes and concluded they had resulted in an annual GHG-emission-reduction of 2.1 million tons—3% of all Swedish GHG emissions (Granberg & Elander, 2007; Kern & Alber, 2009).

Many pioneering cities across Europe and beyond have joined transnational networks in recent years by city networking, a form of *horizontal self-governing* (cf. Kern & Alber, 2009). Municipalities are conducting GHG emission inventories to analyse their reduction potential and measure their performance, evidenced in the initiatives of transnational city networks, such as the International Council for Local Environment Initiatives' Carbon Disclosure Project. In August 2008, around 30 urban centres in the US agreed to use the same methodology to monitor and disclose their emission data. In addition, municipal governments across Europe are in preparation of meeting the goal EU of nearly ZEB. Two important EU initiatives and (financial) support programmes will be presented in the following paragraphs, which support pioneering cities in meeting the goal of the 2008 EU climate and energy package at least a 20% reduction of GHG emissions below 1990 levels by 2020 in developing CO_2 emission inventories, and in ZEB.

3.3. *Covenant of Mayors and Financial Support Programmes*

By signing the "Covenant of Mayors" (EU Mayors, 2010), they can voluntarily commit themselves to reducing their cities' CO_2 emissions by at least 20% by 2020. To meet this aim, and as a prerequisite to elaborate a "Sustainable Energy Action Plan" (SEAP), signatories have to develop a Baseline Emissions Inventory, which provides knowledge of the nature of the entities emitting CO_2 on the municipality's territory. They also need to elaborate a SEAP, which must integrate the built environment, local energy networks and urban transport systems. It is also mandatory for signatories to produce a report every second year to monitor progress towards the self-set targets. One of the signatories is Heidelberg (Germany), which has launched a plan to build a large, zero-emissions city district called "Bahnstadt" on the site of a former goods station (EU Mayors, 2010). Buildings will be constructed according to Passive House standards and renewable energy will be supplied by district heating from biomass thermal power plants and deep geothermics. Another signatory to the Covenant is Amsterdam, which will be treated later in this paper. In the beginning of 2011, more than 2100 cities have signed the Covenant. The European Commission has set up a "Benchmarks of Excellence" system to provide good examples of signatories' initiatives, and has implemented the Covenant's Office, which provides technical and promotional support, including the implementation of tools for evaluation and monitoring, and mechanisms to facilitate sharing of know-how between territories.

There are a number of financial support programmes and initiatives, signatory cities can make use of, such as the European Local ENergy Assistance (ELENA), a technical assistance grant facility, supporting, for example, investments in public and private buildings (including social housing and street and traffic lighting) and district heating/cooling networks, such as CHP and decentralized CHP systems (EIB, 2011).

3.4. *Smart Cities Initiative*

The EU Smart Cities Initiative supports cities and regions in taking measures towards a 40% reduction of GHG emissions (compared with 1990 levels) by 2020 through sustainable energy use and production, providing advantages in terms of quality-of-life, local employment and businesses, and citizen empowerment (EC/SETIS, 2011). It intends to spread across best practices of local sustainable energy concepts across Europe, for instance, through the Covenant. By 2015, local authorities may apply holistic approaches, including "net zero" energy requirements or carbon emissions when averaged over the year for new buildings. Apart from sustainable energy network demonstration and other programmes, the initiative includes the testing of 100 new residential buildings for different design options for ZEBs in different climatic zones: the focus is particularly on design integration of different technologies to prove cost-effective solution and on monitoring the performance under real use (EC/SETIS, 2011).

4. Case Study Results

This section presents the climate policies and instruments used and developed by the municipal governments of Amsterdam (actively working on horizontal self-governing as signatory of the Covenant of Mayors) and Almere to promote new housing projects that surpass the current EPC. It also elaborates two cases: Buiksloterham (Amsterdam) and Zoneiland Almere (Almere). The cases demonstrate, that by the selection of market parties in tenders on the basis of sustainability criteria (Buiksloterham) and a higher EPC (Almere), higher EPC levels as a step towards "energy-neutral" or "nearly zero-energy" building, can be achieved. They also offer perspectives for the promotion of energy-neutral new housing developments in other European countries.

4.1. *Climate Policy Amsterdam and Buiksloterham*

Amsterdam is the capital of the Netherlands. With 767,333 residents (Gemeente Amsterdam, 2011b), the largest city in both the province of North Holland and the country. The city comprises the northern part of the "Randstad", the sixth-largest metropolitan area in Europe, and is divided into seven city parts, each with its own council ('stadsdeelraad").

The Municipality of Amsterdam aims to achieve a 40% reduction in CO_2 emissions by 2025, compared with 1990 levels. The city council has decided that starting from 2015, all new dwellings and commercial and industrial buildings must be constructed as "climate neutral" (Klimaattafel, 2008). Amsterdam's climate policy[5] is established in the "Air Quality Action Plan", "Climate Vision" and "New Amsterdam Climate" programme. The municipality established a "Climate Table" (Klimaattafel) to implement the programme and to set up lasting public–private alliances. A policy note entitled "Sustainability in new buildings" distinguishes two ambition levels (Gemeente Amsterdam, 2009a): (1) climate-neutral with an EPL- (Energy Performance on Location) score between 9 and 9.5, and (2) semi-climate neutral, with an EPL-score of 8. EPL is used to calculate area-bound energy performances with 10 as the maximum score. The Passive House method needs to be applied for building-related energy performances. A municipal working group has developed the following instruments for dwellings (Gemeente Amsterdam, 2009a):

- *A calculation model*, which shows how the climate-neutral building can be achieved for multifamily apartments in Amsterdam by two variants, following the Trias Energetica[6]: one making use of city heat[7] and solar energy, and the other of Passive House building combined with heat-and-cold-storage. This may be achieved with around €13,000 additional costs, expected to be paid back through lower energy bills;
- *A planning and decision-making process for spatial measures*, the so-called "Amsterdam Plaberum" ("Plan- en besluitvormingsproces ruimtelijke maatregelen"), which obliges project owners, such as urban project groups, to draw up an energy vision in the research phase;
- *A tendering procedure, in which sustainability plays a role in the selection of market parties* and in the allocation of building packages, and which fits within the European tendering regulation.

Buiksloterham, an industrial estate in Amsterdam-North will be transformed into a mixed-use urban area with 2000 dwellings to be realized by 2015. Since 1896, the policy for the municipality has been to retain bare ownership and to sell a right of "emphyteusis" ("erfpacht"), in Amsterdam an enduring title, comparable in some aspects to freehold with a restrictive covenant, that may be assigned to private agents (Korthals Altes & Tambach, 2008). The municipality owns a major share of the industrial estate, mostly bare ownership, is acquiring land and rights of emphyteusis, and the land is disposed of in "building packages" with a specific re-development programme for larger lots, to give market agents more scope for optimizing functions within each lot (Korthals Altes & Tambach, 2008).

The municipal ambition is to develop the Buiksloterham climate neutral, and it served as a pilot for the tendering procedure. The municipality issued a first public "sustainability tender" for five lots for which the land price was fixed, and tenderers had to sign for the entire course of the selection process (Gemeente Amsterdam, 2009c). Building costs for dwellings and business accommodations, and the future exploitation are paid by and for the risk of the developer. Prior to the tender, developers were provided all the necessary background information, including an entry brochure with the legal protocol; the "Sustainability in new buildings" note with climate ambitions for both city and area and sustainability definitions; "building packages", including a "sustainability paragraph" with requirements in the field of energy saving, sustainable energy generation, waste, and use of materials, at least 30% plants and trees per lot, by referring to GPR©gebouw and the Amsterdam "calculation table for climate-neutral building" (Gemeente Amsterdam, 2008). GPR©gebouw (with 10 as best score) is used to calculate their environmental burden of buildings, and the calculation table includes an EPL-score, "stepping stones" and scores for the climate-neutral building. The scoring system supports the execution of only the first two steps of the Trias Energetica.

Pre-selection was based on a public tender based on the entry brochure with the following criteria: (1) a "general sustainability vision" (company philosophy and experience in sustainable project development) combined with a "sustainability vision for Buiksloterham" (ambition in the field of project development, technology, consumers (either occupants or entrepreneurs) and future value), (2) the company profile, and (3) the financial situation of the developer. Per lot, a jury selected the four (maximum) most promising tenderers. Pre-selected tenderers participated in a non-public round, in which they submitted their building plans (designs) for the definitive selection, and an evaluation on sustainability aspects (cf. SenterNovem, 2009). The latter had to consist of two proposals: one for a "quality score" in the field of "sustainable building" (by using GPR©gebouw version 4.0), and "climate-neutral building" (using the table, and EPL Quickscan[8] including an EPC value for dwellings), and the other for the design. Together, both scores made up the "score in the field of sustainability", which weighed the highest in the definitive selection. Moreover, plans were assessed on their architectural/urban- development translation of "sustainability" in the design and on their consistency and coherence with the sustainability vision. Both the "score in the field of sustainability" and the "design score" form the definitive score of the plan. The plan with the highest score would be nominated by the selection committee (Gemeente Amsterdam, 2009a, 2009b, 2009c). Candidates for the lots were exempted from the obligation to connect new buildings to the city-heat net if their plans were calculated "climate neutral" through the use of other measures (Timár, 2010).

The committee selected four winners, mainly small developers, cooperating with small architectural firms with innovative, climate-neutral designs (cf. Van Poelgeest, 2010). One winning design (Figure 1(a) – (c)) with PV panels, city heat and cooling is projected on a lot on which at least 35% up to 70% dwellings and at least 30% businesses must be realized. This mixed-use building plan has an average GPR-score of 9.37. The new blocks of flats score 9 for energy, 9.9 for health, 9.8 for future value, 9.7 for usage quality and 8.8 for environment. In comparison, a GPR score of 6 corresponds to the standards set by the Dutch Building Decree (2003) for new construction.

4.2. *Climate Policy Almere and Zoneiland Almere (Solar Island)*

Almere is the seventh largest city in the Netherlands and the largest city in the province of Flevoland, with 190,000 residents and 13,000 businesses (Gemeente Almere, 2011). It has

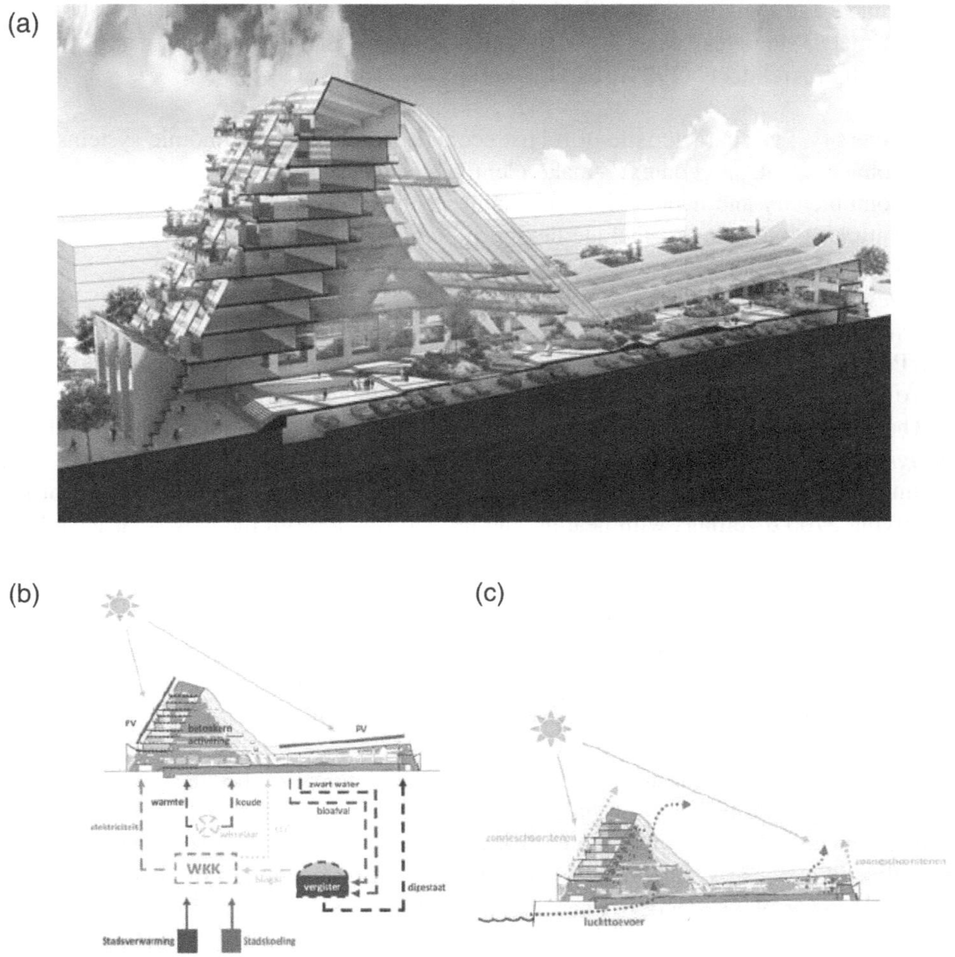

Figure 1. (a–c) Design "De Kameel van Noord" by Heddes Vastgoed, Op ten Noort Blijdenstein architecten en adviseurs, and Deerns raadgevende ingenieurs
Source: Gemeente Amsterdam (Projectbureau Noordwaarts) (2010)

been one of the fastest growing cities in Europe since its establishment in 1976, and has grown into the status of "Holland's most exemplary new town" since its initial development as a suburban area east of Amsterdam (Gemeente Almere, 2011).

Since the 1970s, Almere has committed itself to a programme of sustainable development (Crrescendo, 2007). Almere's 1997 policy note entitled "Sustainable New Living Environments" and the letter of intent "Sustainable Building Flevoland" of 1998 specified the minimum level for sustainable house-building, including a 10% lower EPC standard than required by the Building Decree at that time (Gemeente Almere, 2003). Almere's commitment to the sustainable development of the city is continued in its Environment Plan 2003–2007, which aims at achieving a 20% CO_2 reduction within six years, 25% sustainable energy for housing by 2010 and the inclusion of a large-scale wind-energy network (Crrescendo, 2007). Today, the municipality aims to realize the ecological, economical and socially sustainable development of Almere by 2030. To achieve this goal, the "Almere Principles" were formulated, based on the Hannover Principles by William McDonough Architects (1992) (DLG, 2009, p. 18). These are:

(1) Diversity—as characteristic of robust ecological, social and economic systems
(2) Connect place and context—make identity stronger
(3) Combine city and nature—increase human bond with nature
(4) Anticipate on change—include flexibility in plans and programmes
(5) Keep innovating—in processes, technologies, infrastructures
(6) *Design healthy systems—"cradle to cradle"* (McDonough & Braungart, 2002) *in urban systems*
(7) People make the city—citizens are driving forces behind the making and preservation of the city, as well as behind making it more sustainable

The design of "Zoneiland Almere" (solar island) is based on the sixth principle, that of *designing healthy systems*. The solar island (Figure 2) has a surface of 15,000 m^2 with around 7000 m^2 solar collectors that heat water, which is pumped into the city-heat net, supplying 2700 dwellings with heating and hot tap water. With the new solar island, the Municipality of Almere has created a new icon and entry into the Noorderplassen-West district. In 1997, the municipality issued a tender for the integral energy supply of this district. Requirements were a CO_2 reduction of at least 30%, an EPC-sharpening by 10%, and energy to be sustainably generated for 10%. The 1998 tender by Nuon was selected as the best, including a proposal for connecting the district to the existing city-heat net of Almere Stad, using collective solar collectors, which would renewably generate part of the city heat.

In 2008, the final contract was signed by the municipal authorities of Almere and Nuon. Nuon Warmte has now built an oval island, covered by 520 solar collectors, which generate 9750 GJ per year—10% of the total annual energy demand of 2700 new dwellings. Nearly, all (90%) of this demand will be supplied by waste heat from a nearby power plant. Energy is generated by a heat-power plant and by a solar island, bundled by a heat-transmission station, and distributed via a district heating grid to the households (tap water). The solar island is public–privately financed: about half of the costs are borne by Nuon, about 1/3 by EU subsidy and about 1/6 by some buyers of private lots and of new houses in the district and by the municipal government (detailed information: cf. Gemeenteraad Almere, 2009). Combined with district heating, the solar island's collectors are expected to reduce CO_2-emissions by 50%. At present, 2000 "eco-homes" are being built in the Noorderplassen-West and the Columbus districts. "Specific innovations with these developments include the integration of renewable energy sources and the application of energy efficiency measures in city planning, area development, public tendering, architecture and building. 500 of the new dwellings will be certified as 'solar homes': wood-built, low-energy homes with solar power supply" (Crrescendo, 2007).

5. Discussion

In line with the EPBD recast (EC, 2010), promoting the realization of nearly ZEBs, the Dutch government has expressed the intention to increase the current energy performance standard of the national building regulations for new buildings (EPC) step by step to an

Figure 2. Artist impression solar island
Source: Gemeente Almere, 2010

energy-neutral level by 2020. This research focused on how and through which policy instruments Dutch municipal authorities can promote energy-neutral new housing developments. The case studies in Amsterdam and Almere indicate that municipal authorities can create a framework for energy-neutral new housing developments, if they follow *an integrated design strategy for sustainable energy systems throughout the city: in both urban areas and buildings*.

The municipal government of Amsterdam clearly plays a strong role in governing by enabling (Kern & Alber, 2009), as illustrated by the use of the "Climate Table", which was established to implement the climate programme and to set up, co-ordinate and facilitate public–private alliances. In the "sustainability tender", the municipality disposed of land in "building packages" by posing requirements in the field of sustainability for the development of new residential buildings. However, for other European countries this solution may be limited to municipalities that have a comparable situation regarding land ownership. In this respect, England's leasehold resembles most the Dutch "erfpachtrecht", although there are also lease-forms that are more comparable to a rent (in Dutch: "huur") (cf. De Jong & Ploeger, 2008).

Notwithstanding, the sustainability tender seems to offer a "legally valid" answer to the Article 122 barrier of the Dutch Housing Act. In this tender, the developers were committed to meeting sustainability criteria (rewarded by a higher score) by the municipality, and they had to use the calculation tools GPR©gebouw, EPL and the calculation table for climate-neutral building. The *selection of developers in tendering procedures on the basis of sustainability criteria, including the energy performance of dwellings* shows that an enabling governance mode, in which the municipality uses authoritative powers, can be effective. It therefore adds a new instrument to the classification by Kern and Alber (2009, Table 1). The Municipality of Almere also successfully issued a tender and selected Nuon that best fulfilled its requirements on CO_2 reduction, a sharpening of the EPC and sustainable energy generation. Whereas the best plan is often selected by an EMAT (Economically Most Advantageous Tender) norm (scoring), in which the provider of the best price/quality ratio is chosen (SenterNovem, 2009), in this tender, the sustainable quality of the plan weighed highest in the selection process.

The *encouragement of community engagement* can be regarded necessary for the implementation of both climate mitigation strategies (cf. Coenen & Menkveld, 2002), and climate adaptation strategies (cf. Adger, 2003). This research indicates that it is also necessary for the promotion of energy-neutral new housing developments, which entail the implementation of both mitigation and adaptation (e.g. sustainable cooling) strategies. For example, in Schellewoude (Amsterdam-North) the "Stichting Mijn CO_2 Spoor" corporation, which cooperates with the council of Amsterdam-North, showed Al Gore's movie "An Inconvenient Truth" in a district building (Personal communications of 8 March 2011). Inhabitants who viewed the movie decided to place wind turbines in Amsterdam-North to generate electricity for 20% of its households handing in 600 signatures for the plan at its daily management (Onze Energie, 2011). The council provides full co-operation on enabling the placement of the turbines, and the inhabitants have organized themselves as local sustainable energy company "Onze Amsterdamse Noord Energie Coöperatie U.A." (Onze Energie, 2011).

Local climate policy targets, performance indicators such as EPC and GPR©gebouw, definitions and calculation methods for ZEBs vary per municipality and country, which will make it difficult for municipalities to monitor progress and meet the aim of "nearly ZEB". The need to "speak one language" by using common labels such as BREEAM and LEED to make sustainable buildings and areas measurable in terms of quality and quantity was stipulated by Van Hal (2007). However, Collier (1997) identified the absence of energy-efficiency standards—besides budget constraints imposed on local authorities—as "unfavourable". The Covenant of Mayors includes a methodology for cities to monitor progress and the Smart Cities initiative the testing of new residential buildings for ZEB design options. Regarding the high number of signatories, the Covenant can be seen as a "catalyst in recognizing and supporting local government actions, and has created a need for constructing political and institutional frameworks that will sustain and scale-up support" (Egenhofer *et al.*, 2010). *(Inter)regional cooperation between cities in similar climatic zones on the development of both more uniform calculation methods for ZEBs and of a more uniform system to monitor progress towards the self-set targets* can be supported by the Covenant and the Smart Cities initiative. But as Meeus *et al.* (2011) pointed out, the consistency between the two initiatives must be ensured so that they reinforce each other.

National subsidies for municipalities are financially limited, and not permanently available (cf. Tambach *et al.*, 2010) and they have limited opportunities to generate funding for climate protection measures (Kern & Alber, 2009). The first Dutch UKR subsidy tender supported 13 new housing projects, and the grants for "excellent areas" to experiment with house-building surpassing the current EPC were granted to only 13 areas. The BANS and SLoK subsidy schemes appeared to be financially limited, and not permanently available (cf. Tambach *et al.*, 2010). So are KLIMP and LIP, but they seem effective in cutting GHG emissions in various sectors. Nevertheless, municipal co-financing is a requirement in all four funding schemes.

6. Conclusions and Recommendations

This research suggests the following municipal climate governance approach and instruments for the promotion of energy-neutral new housing developments by Dutch municipal authorities: (1) an integrated design strategy for sustainable energy systems throughout the city: in both urban areas and buildings, (2) market-party-selection in tendering procedures based on sustainability criteria, including the energy performance of dwellings, (3) the encouragement of community engagement for the deployment of renewable energy sources, (4) (inter)regional cooperation between cities in similar climatic zones on the development of both more uniform calculation methods for ZEBs and of a more uniform methodology for cities to monitor progress towards the self-set targets.—However, stable and long-term-oriented European and national financial support for municipalities seems to be needed for municipal climate governance to become successful.

Acknowledgements

This research was conducted within the framework of the EU URBACT II project Housing Praxis for Urban Sustainability (HOPUS), which united academics and practitioners from across Europe. We would like to express our gratitude to the interviewees. The input they provided was of great value for this research.

Notes

1. In the Netherlands, the national Housing Act (1901) refers to the national Building Decree (2003) which contains technical building regulations for both new and existing buildings and for various building functions. It also includes minimum requirements in the areas of safety, health, usability, energy efficiency and the environment.
2. The Dutch government also signed the Energy Saving Covenant Housing Associations (2008) with Aedes, the branch organisation of Dutch housing associations (thereby subscribing to the reduction aims of the Spring Agreement), and with the Nederlandse Woonbond, the federation for Dutch tenants and tenant organizations.
3. Compared to the building-related energy consumption in buildings, built under 2007 building regulation and under scope of Dutch energy performance norm "EPN". The new building production includes housing, commercial and industrial buildings.
4. In the first BANS climate covenant (1999), national and local governments agreed to dedicate themselves to reduce CO_2-emissions.
5. Amsterdam's city parts draw up own climate plans, for example Stadsdeel Amsterdam-North's "Wind in de rug—Klimaatplan 2009–2012" in which the ambitions of the Stadsdeel are formulated with regard to climate governance.
6. The sustainable design strategy 'Trias Energetica' (Duijvestein, 1997) follows three distinctive steps:
   (1) Reduce the energy demand, by taking energy saving technologies;
   (2) Use sustainable energy sources as much as possible;
   (3) If there is still an energy demand left, use fossil fuels as efficient as possible.
   The notion of 'Trias Energica' was introduced in 1996 by Novem (Lysen, 1996), and further developed by TU Delft (C.A.J. Duijvestein)
7. From 2008, connection to the city heat net has been obligatory by Amsterdam's building ordinance (bouwverordening), in case the city heat pipe lies within 40 m distance from the new building (Gemeente Amsterdam, 2011a).
8. EPL Quickscan was developed by SenterNovem.

References

Adger, W. N. (2003) Social capital, collective action, and adaptation to climate change, *Economic Geography*, 79(4), pp. 387–404.

Alsema, E. A., Hoiting, H. & Roth, E. (2009) Stevige ambities, klare taal! Definiëring van doelstellingen en middelen bij energieneutrale, CO_2-neutrale of klimaatneutrale projecten in de gebouwde omgeving, Utrecht: W/E Consultants in commission by PEGO/SenterNovem. Available at http://www.w-e.nl/Bestanden/bestanden/publicaties/Rapport%20definities%20CO2%20en%20energieneutraal%20PeGO.pdf (accessed 4 December 2011).

Bai, X. (2007) Integrating global environmental concerns into urban management: the scale and readiness arguments, *Journal of Industrial Ecology*, 11(2), pp. 15–29.

Betsill, M. & Bulkeley, H. (2006) Cities and the multilevel governance of global climate change, *Global Governance*, 12(2), pp. 141–159.

Bulkeley, H. & Kern, K. (2006) Local government and the governing of climate change in Germany and the UK, *Urban Studies*, 43(12), pp. 2237–2259.

Bulkeley, H., Schroeder, H., Janda, K., Zhao, J., Armstrong, A., Chu, S. Y. & Ghosh, S. (2009) *Cities and climate change: The role of institutions, governance and urban planning*, Report prepared for the World Bank's 2009 Urban Research Symposium on Cities and Climate Change, Durham University and University of Oxford, UK, 92 p.

Burger, H., Coenen, F., Kaal, M. & Menkveld, M. (2001) *Evaluatie van het speelveld van lokaal klimaatbeleid*, Rapport Fase 2 NOP-project "Lokale overheden en klimaatbeleid", Petten.

CBS (2011) *Woningbouw: Voorraad, wijzigingen en nieuwbouw naar voortgang, per regio (Statistics Netherlands)*. Available at http://statline.cbs.nl/StatWeb/publication/?DM=SLNL&PA=7413&D1=0-3&D2=0,5-16&D3=7,12,19-l&HDR=T,G2&STB=G1&VW=T (accessed 4 December 2011).

Coenen, F. & Menkveld, M. (2002) The role of local authorities in a transition towards a climate-neutral society (Chapter 6), in: M. Kok, W. Vermeulen, A. Faaij & D. De Jager (Eds) *Global Warming & Social Innovation. The Challenge of a Climate-neutral Society*, pp. 107–125 (London: Earthscan Publishing Ltd).

Collier, U. (1997) Local authorities and climate protection in the European Union: Putting subsidiarity into practice, *Local Environment*, 2(1), pp. 39–56.

Crrescendo (2007) *Increasing the use of renewable energy sources*, p. 20. Available at www.crrescendo.net/pdf/crescendo%20brochure.pdf (accessed 4 December 2011).

Davoudi, S., Crawford, J. & Mehmood, A. (Eds) (2009) *Planning for Climate Change: Strategies for Mitigation and Adaptation for Spatial Planners* (London: Earthscan).

De Jong, J. & Ploeger, H. D. (2008) *Erfpacht en opstal (B28)* ((Deventer: Derde druk, Monografieën BW, Kluwer).

DLG (Dienst Landelijk Gebied) (2009) *Cradle to cradle als inspiratiebron voor duurzame gebiedsontwikkeling*. Available at www.minlnv.nl/cdlpub/servlet/CDLServlet?p_file_id=35183. (accessed 4 December 2011)

Duijvestein, C. A. J. (1997) *Ecologisch bouwen* (in Dutch) (Delft: TU Delft, Faculteit Bouwkunde SOM/BOOM: 59).

EC (European Commission) (2003) Directive 2002/91/EC of the European Parliament and of the Council of 16 December 2002 on the energy performance of buildings, *Official Journal of the European Communities*, L1, 4 January 2003, pp. 65–71.

EC (European Commission) (2010) Directive 2010/31/EU of the European Parliament and of the Council of 19 May 2010 on the energy performance of buildings (recast), *Official Journal of the European Communities*, L 153, 18 June 2010, pp. 13–35.

Egenhofer, C., Alessi, M., Núñez Ferrer, J. & Tubiana, L. (2010) *Greening EU Cities: The Emerging EU Strategy to Address Climate Change* (Brussels: Centre for European Policy Studies).

EIB (European Investment Bank) (2011) *European Local ENergy Assistance (ELENA)*. Available at www.eib.org/products/technical_assistance/elena/index.htm (accessed 23 February 2011).

EU Mayors (2010) *Covenant of Mayors*. Available at www.eumayors.eu/home_en.htm (accessed 4 February 2011).

EC/SETIS (European Commission/Strategic Energy Technology Information System) (2011) *European Initiative on Smart Cities*. Available at http://setis.ec.europa.eu/about-setis/technology-roadmap/european-initiative-on-smart-cities (accessed 23 February 2011).

Fieten, R. M. (2008) Duurzaam bouwen en het Bouwbesluit; de (her)ontdekking van art. 7 en 7a van de Woningwet Nr. 65, *Bouwrecht*, 45(5), pp. 295–299.

Gemeenteraad Almere (2009) *Toeslag duurzame energie Noorderplassen-West*. Available at: http://gemeenteraad.almere.nl/raadsinformatie/schriftelijke_vragen/schrvragen_content/BNschrvragen/_rp_kolom1-1_elementId/1_514725 (accessed 9 April 2009).

Gemeente Almere (2003) *Duurzaam bouwen kan in Almere! Persbericht Gemeente Almere*. Available at: www.ikcro.nl/artikelen/Almere1103.pdf (accessed 4 December 2011).

Gemeente Almere (2010) *Zoneiland Almere*. Available at www.almere.nl/de_stad/stadsdelen/content/_rp_kolom1-1_elementId/1_2243768 (accessed 30 March 2010).

Gemeente Almere (2011) *The City of Almere*. Available at http://english.almere.nl/ (accessed 22 February 2011).

Gemeente Amsterdam (2008) *Duurzaamheid in de nieuwbouw*, pp. 1–14. Available at http://www.amsterdam.nl/publish/pages/171030/nieuwbouw.pdf (accessed 4 December 2011).

Gemeente Amsterdam (Dienst Milieu en Bouwtoezicht) (2009a) *Bouwbrief Regels en afspraken Amsterdamse woningbouw*, 101, pp. 1–6. Available at http://www.oga.amsterdam.nl/bijlagen/bouwbrief/119.pdf (accessed 4 December 2011)

Gemeente Amsterdam (2009b) *Notitie Definities, maatlatten en selectiecriteria bij gunning van klimaatneutrale nieuwbouw*. Available at: http://www.amsterdam.nl/publish/pages/240122/20090304definitiemaatlatselectie.pdf (accessed 4 December 2011).

Gemeente Amsterdam (Projectbureau Noordwaarts) (2009c) *Inschrijfbrochures Voor- en Definitieve Selectie Duurzaamheidstender Buiksloterham. Bouwenveloppen 12, 21, 22, 24, 41*. Available at http://www.noordwaarts.nl/publish/pages/146165/inschrijfbrochuredefinitieveselectieduurzaamheidtenderbuiksloterham.pdf (accessed on 4 December 2011).

Gemeente Amsterdam (Projectbureau Noordwaarts) (2010) *Winnaars bekend duurzaamheidstender BSH*. Available at www.noordwaarts.nl/@297650/winnaars_bekend/ (accessed 17 February 2011).

Gemeente Amsterdam (2011a) *Bouwverordening Amsterdam 2003* (geconsolideerde versie, geldig vanaf 1-10-2010). Available at www.regelgeving.amsterdam.nl/centralestad/bouwverordening_2003 (accessed 8 March 2011).

Gemeente Amsterdam (2011b) (Dienst Onderzoek en Statistiek) *Bevolkingsgegevens derde kwartaal 2009*. Available at www.os.amsterdam.nl/pdf/2009_bevolking_amsterdam_3.pdf (accessed 22 February 2011).

Granberg, M. & Elander, I. (2007) Local governance and climate change: Reflections on the Swedish experience, *Local Environment*, 12(5), pp. 537–548.

Hoekstra, J. J. & Van Der Veen, G. A. (2008) Kwaliteits- en bindingseisen bij nieuwbouw, wat mag wel en wat niet, *Tijdschrift voor Bouwrecht*, (5), pp. 430–442.

Janicke, M. & Weidner, H. (Eds) (1997) *National Environmental Policies: A Comparative Study of Capacity Building* (Berlin: Springer).

Kern, K. & Alber, G. (2009) Governing climate change in cities: Modes of urban climate governance in multi-level systems, in: *Competitive Cities and Climate Change, OECD Conference Proceedings*, Chap. 8, pp. 171–196, p. 175 (Table 1 titled "Modes of urban climate governance and sectors of climate change mitigation"), October 910 2008, Milan, Italy, OECD. Available at http://www.oecd.org/dataoecd/12/38/42554913.pdf (accessed 4 December 2011).

Klimaattafel (2008) *Advies van de Klimaattafel NUON in het kader van net Klimaatprogramma van Amsterdam*. Available at www.dgbc.nl/images/uploads/klimaattafeladviesdefinitief1.pdf (accessed 4 December 2011).

Korthals Altes, W. K. & Tambach, M. (2008) Municipal strategies for introducing housing on industrial estates as part of compact-city policies in the Netherlands, *Cities*, 25(4), pp. 218–229.

Lafferty, W. M. & Eckerberg, K. (1998) *From Earth Summit to Local Agenda 21. Working Towards Sustainable Development* (London: Earthscan).

Lim, B., Spanger-Siegfried, E., Burton, I., Malone, E. L. & Huq, S. (2005) *Adaptation Policy Frameworks for Climate Change: Developing Strategies, Policies and Measures*. Available at http://www.preventionweb.net/files/7995_APF.pdf (accessed 4 December 2011).

Lysen, E. H. (1996) The Trias Energica: Solar energy strategies for developing countries, Proceedings of Eurosun Conference, September 16–19, Freiburg, Available at www.juccce.com/documents/Solar%20Energy/TriasEnergica_Lysem.pdf, URL of JUCCCE or Joint US-China Collaboration on Clean Energy (accessed 4 December 2011).

Marszal, A. J., Bourrelle, J. S., Nieminen, J., Berggren, B., Gustavsen, A., Heiselberg, P. & Wall, M. (2010) North European understanding of zero energy/emission buildings, Zero Emission Buildings—Proceedings of the Renewable Energy Conference 2010, Trondheim, Norway, Available at www.tapironline.no/last-ned/288 (accessed 4 December 2011).

Marszal, A. J., Heiselberg, P., Bourelle, J. S., Musall, E., Voss, K., Sartori, I. & Napolitano, A. (2011) Zero energy building—a review of definitions and calculation methodologies, *Energy and Buildings*, 43(4), pp. 971–979.

McDonough, W. & Braungart, M. (2002) *Cradle to Cradle Remaking the Way We Make Things*, pp. 1–193 (New York: North Point Press).

Meeus, L., Delarue, E., Azevedo, I., Glachant, J.-M., Leal, V. & De Oliveira Fernandes, E. (2011) *Smart Cities Initiative: How to Foster a Quick Transition towards Local Sustainable Energy Systems*, THINK, Final Report. Available at http://www.eui.eu/Projects/THINK/Documents/THINKsmartcitiesReport.pdf (accessed 4 December 2011).

MVROM (Ministry of Housing, Spatial Planning and the Environment) (2007) *New Energy for Climate Policy: The "Clean and Efficient" Programme*, The Hague, 64 p. Available at http://www.rijksoverheid.nl/documenten-en-publicaties/brochures/2010/11/25/new-energy-for-climate-policy.html (accessed 4 December 2011).

MVROM (Ministry of Housing, Spatial Planning and the Environment) (2008) *Kabinetsbrede aanpak duurzame ontwikkeling*. DGM/BREM2008050615, pp. 1–11. Available at http://www.rijksoverheid.nl/documenten-en-publicaties/kamerstukken/2008/05/16/kabinetsbrede-aanpak-duurzame-ontwikkeling-kamerbrief.html (accessed 4 December 2011)

OECD (1995) *OECD Environmental Performance Reviews: The Netherlands* 1995, 232 pp. (Paris Cedex, France: OECD Publications Service).

Onze Energie (2011) *Hoe onze energie begon*, Available at www.onzeenergie.nl/p/1/82/ms1-15/hoe-het-begon (accessed 4 December 2011).

PeGO (Platform Energie Transitie Gebouwde Omgeving) Werkgroep Innovatie (2007) *Energietransitie-Plan (Energy Transition Plan) PeGO*, pp. 1–16, p. 42 (Boxtel: Aeneas).

Schepers, G. (2009) *Klimaatprogramma 2009–2013: Uitvoeringsprogramma Gemeente Nieuwegein*. Available at www.vng.nl/Praktijkvoorbeelden/RWMV/2009/klimaatprogramma_2009-2013_Nieuwegein.pdf (accessed 4 December 2011).

Schroeder, H. & Bulkeley, H. (2009) Global cities and the governance of climate change: What is the role of law in cities? *Fordham Urban Law Journal*, XXXVI(2), Cooper-Walsh Colloquium, pp. 313–359. Available at http://law2.fordham.edu/publications/articles/400flspub17715.pdf (accessed 4 December 2011).

SenterNovem (2008) *Handleiding Tender "UKR Naar energieneutraal wonen"*. Available at http://regelingen.agentschapnl.nl/content/handleiding-tender-ukr-naar-energieneutraal-wonen (accessed 30 March 2010).

SenterNovem (2009) *Duurzaam inkopen versie 2.0. Een model voor MVO-gedreven inkoopbeleid*, pp. 30–35, November 2009, Utrecht. Available at www.rijksoverheid.nl/documenten-en-publicaties/brochures/2009/11/01/duurzaam-inkopen-versie-2-0-een-model-voor-mvo-gedreven-inkoopbeleid.html (accessed 4 December 2011).

Tambach, M. & Visscher, H. (2010) Towards energy-neutral new housing developments: Municipal policy instruments in the Netherlands, in: C. Clemente & F. De Matteis (Eds) *Housing for Europe, Strategies for Quality in Urban Space, Excellence in Design, Performance in Building* (Rome: Tipographia Del Genio Civile.), Available at http://urbact.eu/fileadmin/Projects/HOPUS/outputs_media/hopus_final_LOW.pdf (accessed 4 December 2011).

Tambach, M., Hasselaar, E. & Itard, L. (2010) Assessment of current Dutch energy transition policy instruments for the existing housing stock, *Energy Policy*, 38(2), pp. 981–996.

Timár, E. (2010) *Hybride energieaanpak in Amsterdam-Noord.* Commissioned by Agentschap NL (part of Agentschap NL's "Lessons Learned" series) Utrecht. Available at www.noordwaarts.nl/publish/pages/142749/senternovemlessonslearnedamsterdamnoorddef.pdf (accessed 8 March 2011).

Todaro, B., Clemente, C., De Andreis, L., De Matteis, F., Visscher, H. J., Tambach, M., Almeida, M., Lourenço, J., Mateus, R., Lorens, P., Kabronska, J., Rembarz, G., Kamrowska-Zaluska, D., Milardi, M., Pennestrš, D., Porcino, D., Arnold, J., Van Vliet, A., Kemp, D. & Lamia, J. A. L. (2008a) *Housing Praxis for Urban Sustainability Baseline Study*, pp. 38–45 (TU Delft), Available at http://w3.uniroma1.it/urbact/index.php/project-output/4-project-output (accessed 4 December 2011).

Todaro, B., Clemente, C., De Andreis, L., De Matteis, F., Visscher, H.J., Tambach, M., Almeida, M., Lourenço, J., Mateus, R., Lorens, P., Kabronska, J., Rembarz, G., Kamrowska-Zaluska, D., Milardi, M., Pennestrì, D., Porcino, D., Arnold, J., Van Vliet, A., Kemp, D. & Lamia, J. A. L. (2008b) URBACT II - Working Group HOPUS Housing Praxis for Urban Sustainability, in: De Matteis (Ed.) *Good, Green, Safe, Affordable Housing*, 91 pp. (Verona: Ipertesto Edizioni).

Van de Griendt, J. S. (2010) Omgaan met milieu en duurzaamheid bij vastgoed- en gebiedsontwikkeling (Chapter A7), in: Ed. Nozeman *et al.* (Eds) *Handboek Projectontwikkeling, een veelzijdig vak in een dynamische omgeving,* 2nd revised ed. (Doetinchem: Reed Business Publishers).

Van der Laan, E. E. (2009a) *Brief "Excellente gebieden innovatieve energiebesparing in de nieuwbouw"*, Kenmerk SB 2009066621. Available at www.lente-akkoord.nl/wp-content/uploads/2009/12/Brief-Excellente-gebieden.pdf (accessed 30 March 2010).

Van der Laan, E. E. (2009b) *Bijlage bij brief over "excellente gebieden innovatieve energiebesparing in de nieuwbouw"*, kenmerk SB2009066621. Available at www.lente-akkoord.nl/wp-content/uploads/2009/12/Bijlage-bij-brief-excellente-gebieden.pdf (accessed 30 March 2010).

Van Hal, A. (2007) A labeling system as stepping stone for incentives related to the profitability of sustainable housing, *Journal of Housing and the Built Environment*, (22), pp. 393–408.

Van Poelgeest (2010) Wijk van de toekomst—in Buiksloterham! *Het Parool*, February 17, pp. 14, 15.

VNG ("Association of Netherlands Municipalities") and Ministries of: Housing, Spatial Planning and the Environment, Finance, Agriculture, Nature and Food Quality, Economic Affairs and Foreign Affairs (2007) *Climate Agreement Municipalities and National Government 2007–2011. Working Together on a Climate Proof and Sustainable Netherlands* (in Dutch), Samenwerken aan een klimaatbestendig en duurzaam Nederland Klimaatakkoord Gemeenten en Rijk 2007–2011, agreed upon and signed in The Hague, November 12 2007. Avaliable at www.milieuzorgoverheden.nl/downloads/Klimaatakkoord.pdf (accessed 4 December 2011).

Improving Energy Efficiency of Social Housing Areas: A Case Study of a Retrofit Achieving an "A" Energy Performance Rating in the UK

MINNA SUNIKKA-BLANK*, JUN CHEN*, JUDITH BRITNELL** & DIMITRA DANTSIOU**

*Department of Architecture, University of Cambridge, Cambridge, UK, **The Oxford Institute for Sustainable Development, Oxford Brookes University, Oxford, UK

ABSTRACT *Currently, the majority of the European housing stock falls towards the bottom of the energy efficiency rating scale on the EU Energy Performance Certificate. If governments and businesses are to successfully address ambitious* CO_2 *reduction targets, then it will be imperative that energy-efficient measures and policies focus on existing housing. In order to understand what kind of retrofit is needed to achieve an "A" energy performance rating in social housing, the paper reports the findings of an on-going research project in the UK. The paper draws on a case study from the Technology Strategy Board's "Retrofit for the Future" competition entry in Cambridge. The upgrade strategy improved the home's energy performance rating to A, aimed to radically reduce carbon dioxide emissions (17 kg* m^{-2} *year*$^{-1}$*) and provided affordable warmth for the tenants. In order to get an impression of the actual energy consumption in the case study, energy use behaviour of the household was observed. Based on the barriers identified in the case study, the feasibility of the current UK policy strategies (e.g. Smart Meters and Feed-in-Tariffs) to facilitate the acceptance of energy measures in social housing is discussed.*

1. Introduction

The UK government strategy "Warm homes, greener homes" aims to cut household carbon emissions by 29% by 2020. By 2015, every household should have loft and cavity wall insulation, and by 2020, up to 7 million homes should have more substantial improvements such as solid wall insulation or renewable energy technologies (HM Government, 2010). If governments and businesses are to successfully address ambitious

CO_2 reduction targets, then it will be imperative that planning and energy policies address the existing housing stock (Boardman *et al.*, 2005; Sunikka, 2006; Meijer *et al.*, 2009).

Understanding how an energy-efficient retrofit works in practice is vital in lowering the overall household energy consumption. This paper investigates how an "A" energy performance rating can be reached in a retrofit, considering energy efficiency measures, the use of solar energy and the characteristics of household energy use behaviour. The paper is not limited to energy simulations, but draws on real data from the winning entry to the "Retrofit for the Future", an on-going project funded by Technology Strategy Board (TSB) in 2009/2010. The team consisted of PRP Architects, Hill Partnerships and the Cambridge City Council and Cambridge University. The paper aims to answer the following research questions.

1. How is the level "A" energy rating performance achieved in a social housing retrofit in the UK case study?
2. What are the characteristics of household energy use in the retrofit case study?
3. How do policy instruments such as Smart Meters and Feed-in-Tariffs (FIT) facilitate energy savings in the social housing sector in the UK?

Section 2 presents the methodology. Technical measures and the retrofit process are described in Section 3. Section 4 discusses the characteristics of the household's energy use behaviour in the case study. Section 5 focuses on the current UK policy strategies and their effectiveness in social housing. Conclusions are drawn in Section 6.

2. Methodology

Due to the diversity of the housing stock, barriers to energy saving differ per tenure (Sunikka, 2006; Meijer *et al.*, 2009). This research focuses on social housing where economic constraints are dominant, but the stock managed by professional landlords making the decision-making process easier than in the private rental sector. The UK social housing sector provides affordable housing for households with an average income of less than £11,000 a year. In total, a 29% reduction in the emissions from 2008 levels is expected from the UK social housing sector by 2020 (Communities and Local Government, 2010). Saving targets are driven by the fact that around 20% of houses in major regions in the UK is estimated to suffer from energy poverty (Pollitt, 2010). The new "Warm Home" standard aims to raise the energy efficiency standard of social housing from around Standard Assessment Procedure (SAP) 59 to at least 70 (HM Government, 2010)—if the new coalition government will follow the policy. The SAP is adopted by Government as the UK methodology for calculating the energy performance and CO_2 emissions of dwellings (BRE, 2009). Based on the floor area, volume, infiltration, opening areas and *U*-value of different elements, the SAP calculates the heat losses, internal gains, lighting and water heating demand. This information is turned into an energy consumption figure, considering the information provided about the ventilation, space and water heating systems. Building orientation and obstructions to sunlight are used to calculate solar gains and contribution of micro-renewables. Any considerations about building occupancy and use are in proportion to the dwelling area and cannot be changed. In order to overcome financial barriers, a "Green Deal" approach will be introduced to financing upgrades; energy companies will be obliged to support householders in energy saving, including subsidies and free upgrades for households identified as vulnerable (HM Government, 2010).

The case study is located in a social housing area at the southwest side of Cambridge city, in Trumpington. The social housing scheme, which today is partly owned by the Cambridge City Council, was built in the 1950s by the British Steel Homes company as part of the large housing development. The case study (Section 3) and most houses in the area (participating in the survey in Section 4) are all identical three-bed semi-detached properties (86 m^2 floor area), constructed using a steel frame (BISF) structure. Most of the households receive housing benefits, are considered as "fuel-poor" (paying more than 10% of their household income for energy) and, due to their economic situation, are pre-pay metre (PPM) clients, with high energy tariffs than in a traditional contract. As a part of the TSB "Retrofit for the Future" competition, the case study was retrofitted and monitored as a prototype for other social housing in the area.

The retrofit strategy had to be feasible and replicable on a mass scale to any house of this type in any location in the UK. In total, 36,000 BISF houses and 1500 steel-framed Howard Houses were built after the war in the UK. A more ambitious Passivhaus standard, where the heating demand would be reduced to <15 kWh m^{-2} $year^{-1}$, would have required mechanical ventilation, heat recovery and higher insulation levels in the case study—especially in walls and floor—and the elimination of all cold bridges. Despite the mechanical ventilation, there were doubts whether a tight Passivhaus envelope was suitable as the couple smoke indoors and have pets. Furthermore, according to Galvin (2010), the higher the thermal renovation standard, the higher the cost in terms of money invested per unit of energy saved; the Passivhaus standard being less cost-effective compared with a normal retrofit that complies with the building regulations.

Occupant behaviour and household characteristics have emerged as important contributors to domestic energy consumption (Liao & Chang, 2002). Heating energy saving achieved through retrofit measures can be remarkably lower than calculated (Haas & Biermayr, 2000). There seems to be up to 50% difference in energy consumption between the prediction and the reality in a retrofit—and very little knowledge on the reasons why. In the case study, energy use behaviour of the tenants was observed in order to get a picture of the actual energy performance after the retrofit (Section 4). The observations on energy use behaviour are based on energy measurements and comfort surveys in the case study house before and after sustainable retrofit. This is put into the context of a survey on energy use patterns in the identical social housing in the area (13 households).

According to Uitdenbogerd *et al.* (2007), the most important personal (internal) determinants that relate to energy use are the age of occupants, type and age of the house, household type, education, income and the employment situation. External determinants are related to the context: access to financial resources and services, price measures and regulatory measures. The case study focuses on personal determinants. In the context of this study, it was not possible to carry out a large sample on energy use behaviour, but the observations are used to understand the actual energy performance of the building and to indicate areas for further research.

Energy and planning policies tend to focus on investment behaviour, whereas what happens after the energy measures are adopted is often ignored (IEA, 2008; Bergman *et al.*, 2009; Tambach *et al.*, 2010). Policy discussion (Section 5) follows the recommendation of Egmond and Bruel (2007) who state that instead of the current "instrument orientation", policies should be drawn from the identification and removal of barriers (identified in Sections 3 and 4). It should be considered that the policy developments described in this paper are subject to change and based on the situation in 2009/2010.

3. Technical Measures and the Retrofit Process

The retrofit strategy provided substantial improvements to the thermal performance of the external fabric of the house, changes to the building services and the introduction of micro-renewables (Table 1 and Figure 1). It improved the case study's energy performance rating (Energy Performance Certificate, EPC) to "A" and aimed to reduce the annual carbon dioxide emissions to below 17 kg m^{-2} $year^{-1}$ and primary energy consumption to below 115 kWh m^{-2} $year^{-1}$. EPC's are part of the UK's government commitment to the European Energy Performance of Buildings Directive. EPBD2002/91/EC came into force on 4 January 2003. They provide from "A" to "G" energy efficiency rating of a building, "A" being the best performance. Since October 2008, an EPC is now required in the UK for all homes, commercial and public buildings when bought, sold, built or rented. An air permeability value of 5 m^3 h^{-1} m^{-2} was targeted after a pressurization test was conducted, according to which the measured air permeability of the dwelling was 13.58 m^3 h^{-1} m^{-2} under a pressure of 50 Pa. In order to achieve a tighter building envelope, and an "A"-rated performance for the case study house, the following measures were considered necessary and applied (Figures 2 and 3).

3.1 *Building Fabric*

The existing external render in the ground floor and cladding in the first floor were removed, revealing the steel structure. High levels of insulation were added to the walls (over the vapour barrier), and the walls were re-rendered. Additional insulation was placed in the roof and the parts of the ground floor, which were not suspended and not of solid construction. About 200 mm of phenolic foam was applied to the walls (*U*-value of 0.12 W m^{-2} K^{-1}) and 350 mm of mineral wool in the roof (0.12 W m^{-2} K^{-1}). About 25mm of Aeorogel was placed under the floorboards in the hall, the living room and the dining room (0.43 W m^{-2} K^{-1}).

Table 1. Technical information about the property prior to the retrofit

Build date	1947
Floor area (m^2)	86
Construction	BISF (British Steel Association) pre-fabricated
Roof type	Pitched DECRA roof (renovated 2001)
Loft insulation (mm)	150
Wall type	Metal sheet cladding (1st floor) and painted render (ground floor)
Front and back door	Timber (installed 1992/front and 1996/back)
Windows	PVCu double glazing (installed 1997)
Heating system	Vaillant Ecotec Plus 831 Gas Condensing Combi boiler, A-rated (installed in 2009)
Retrofit team	PRP Architects, Hill Partnerships, Cambridge City Council, Cambridge University, Department of Architecture

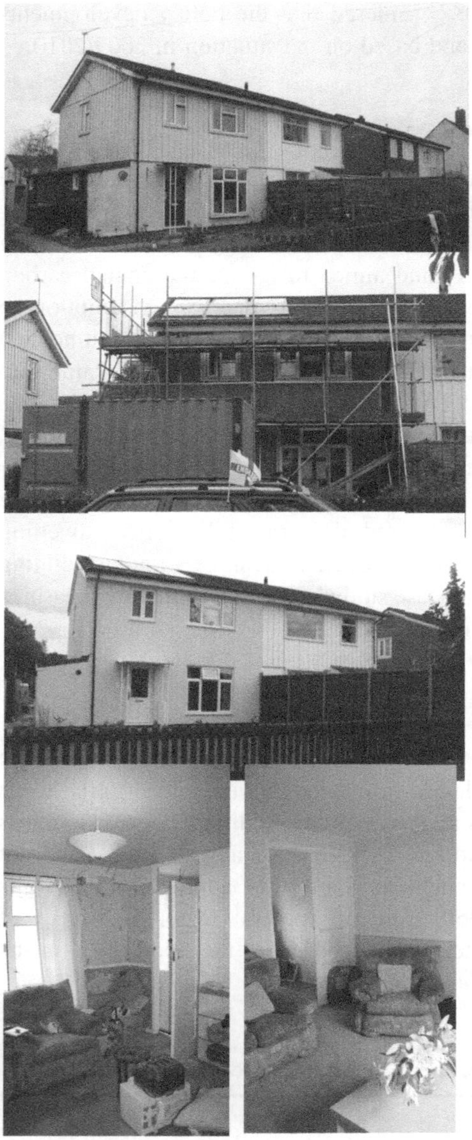

The case study house in Trumpington before the retrofit (09.03.2010)

The case study during the construction (14.06.2010)

The house during the open day (15.07.2010)

The living room during and after the retrofit (14.06.2010 and 15.07.2010)

Figure 1. Site photographs of the retrofit case study in Trumpington, Cambridge

The existing double-glazed windows were replaced with the triple-glazed ones with rigid PVC window frames (U-value of 1.1 W m^{-2} K^{-1}), in order to align the windows with the new cladding and to reduce cold bridges. The external doors were also replaced with triple-glazed (U-value of 1.5 W m^{-2} K^{-1}) security doors.

All materials and component junctions were carefully detailed at the design stage to minimize thermal bridges and to ensure the continuity of insulation. The steel frame was adjusted locally to allow new windows to be supported in line with the new cladding, which continues below the ground level to protect the concrete slab.

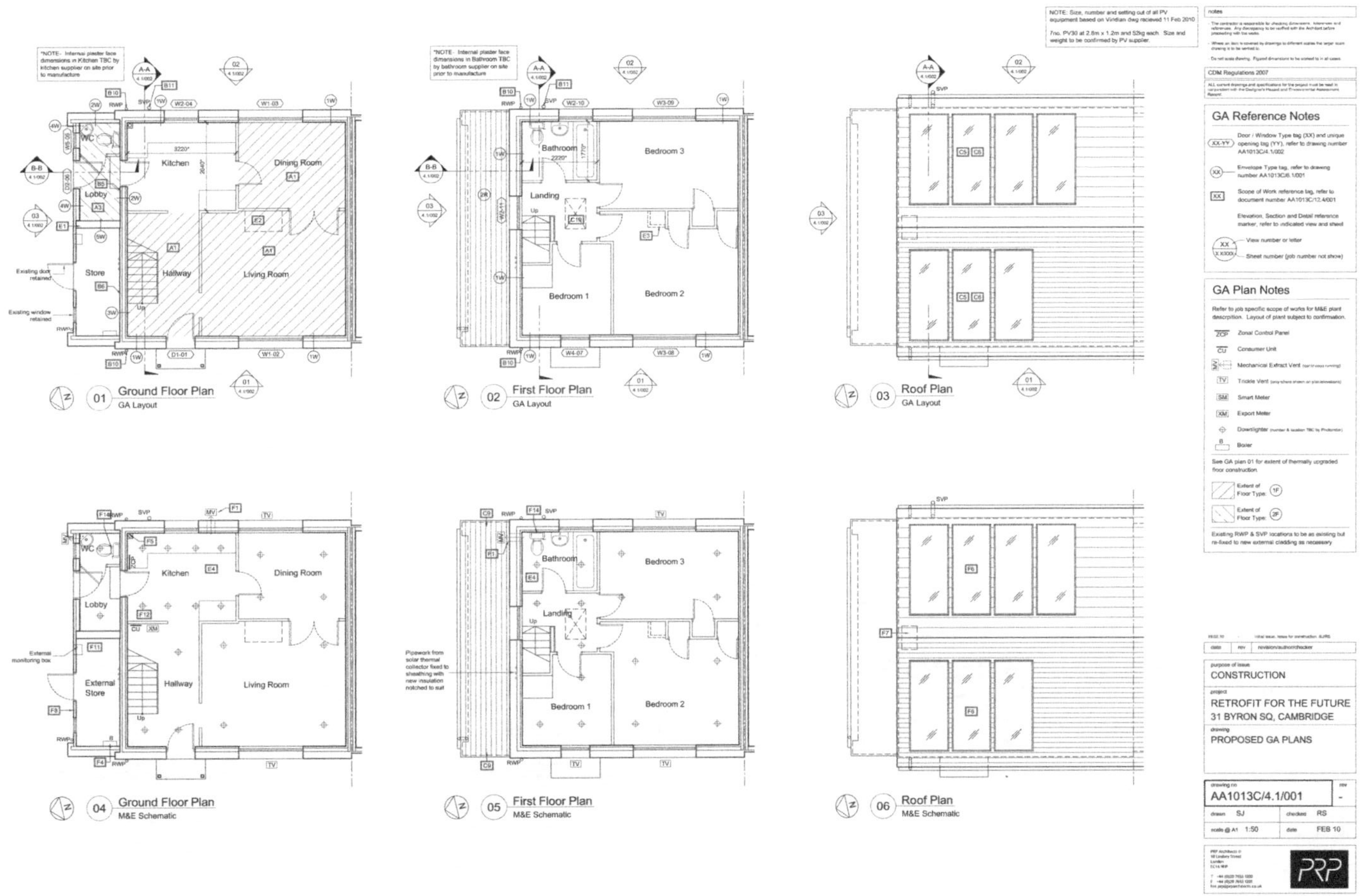

Figure 2. Proposed plans for the retrofit house

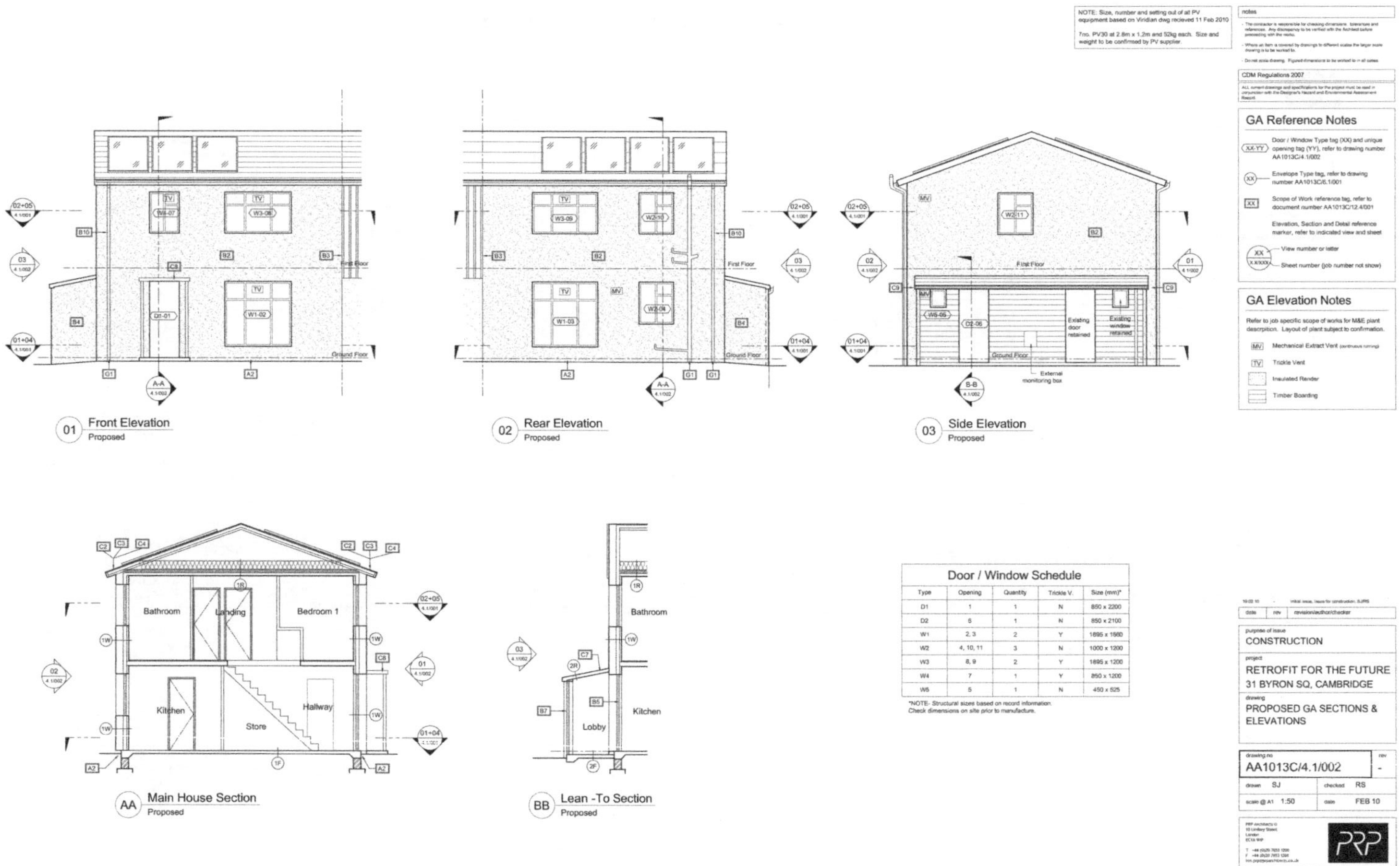

Figure 3. Proposed sections and elevations for the retrofit house

3.2 *Building Services and Micro-renewables*

In order to achieve an "A" energy performance rating, the following installations were added:

- a high-efficiency, low-emission balanced flue gas boiler (with a flue gas heat recovery) to serve the existing radiators;
- a waste water heat recovery unit to the bath and shower;
- 3 m^2 of solar collectors to provide solar thermal;
- 22.5 m^2 of polycrystalline photovoltaic (PV) systems to generate electricity;
- a mechanical extract ventilation (continuously running) with a user-controlled boost in the kitchen and the bathroom and trickle ventilators in the new windows;
- new recessed LED lighting, which required a considerable amount of work in ceilings throughout the house;
- an energy display system including an interactive touch screen interface to educate and increase the awareness of the tenants;
- "A++" energy-efficient white goods in the kitchen.

The retrofit includes an installation of seven polycrystalline PV panels with a surface area of 3.3 m^2 each (1.2 m×2.8 m) and a rated output of 2.7 kWp. The PVs were placed in the optimum tilt of 32° and 20°, respectively, with no shading from the surroundings. The case study house has a west-east-oriented roof, estimated to achieve an annual PV yield of 1876 kWh, depending on the tenants' energy use behaviour and an effective load matching. In order to understand the scale of behavioural factors, two scenarios of electricity consumption were modelled with the "IES-VE" software. A scenario of an occupancy with an unemployed family with three young children (as in the case study) is compared with a working couple, including factors such as the number of occupants, the age and occupation, the heating and lighting schedule and the quantity and the use of appliances. Figure 4 shows that the differences between consumption patterns of the two household types are significant: a working couple can, theoretically, cover their energy needs with the electricity generated by the PVs fully during 5 months of the highest solar radiation (from April until August) and provide

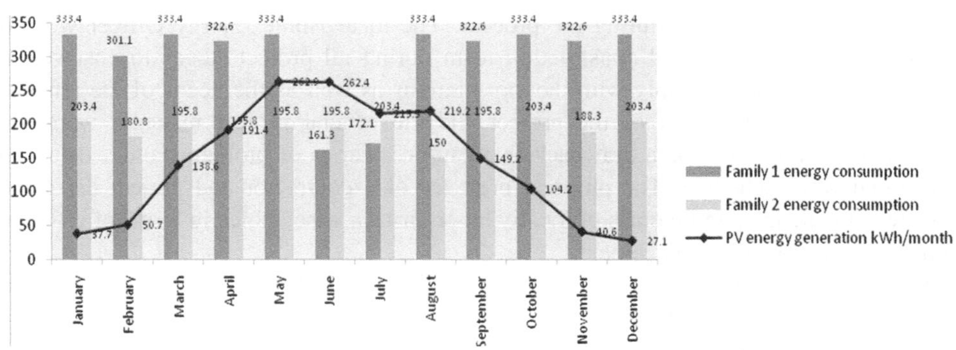

Figure 4. Comparison of electricity consumption and PV energy generation in two different household types (family 1 is a working couple and family 2 an unemployed family with three kids as in the case study)

an electricity autonomy of 82%, whereas an unemployed couple with a family can cover only 52.6% of their total electricity demand with the PVs, even during the summer.

After the installation of the PV panels, electricity generation was closely observed for 2 weeks in July 2010. According to the actual PV meter readings, the system had a mean daily generation rate of 5.6 kWh and a monthly output of 175 kWh. The electricity generation for July was 23% lower than that assumed by the simulation software, displaying a considerable difference between the projected and actual supply.

3.3 *Process*

The case study household was chosen in a competition held by the Cambridge City Council between the eligible houses for the project. According to the Home Energy Officer, the Council wanted a family who would be motivated, engaged with the retrofit and benefit from the savings in the energy bills. The construction period took 9 weeks. The construction team consisted of 12 workers and a site manager, arriving at 8.00 am and leaving at 3.30 pm daily. The tenants and their two youngest children were living in the house during the whole process. For a period of 3 weeks, the work would simultaneously occur in several places, in the kitchen, the living room and the bathroom, leaving little space for tenants and restricting them to one room. According to the tenants, the retrofit was "difficult" and "disruptive, especially after the first two weeks when the work started inside the house". Main inconveniences were a lack of private space for the family and the presence of the youngest child on the site, especially when the noise levels were high. The tenants would prefer to have been moved somewhere else during the weeks that intensive work was taking place in the house interior ("it is very important for the family either to move out during this period or to have a caravan in the garden"); this could also provide more space for workers and accelerate the construction process. Careful consultation and detailed discussions with occupiers would need to take place at the start of any such project in the future, setting out clear boundaries for both parties.

The parties involved in the project comprised the property owner (the Cambridge City Council), who initialized the project, the design team and the contractor. This was the first time a retrofit project of this scale was attempted by the owner, and the lack of experience became apparent at times during the process. The local Home Energy Officer worked closely with the design and construction team during all project stages to ensure that the final result would comply with the expectations as well as the needs of the tenants. The fact that the architect of the project and the official project manager were not locally based proved to be a drawback that caused a series of problems and a delay on the initial work schedule. The physical presence of a contractor and project designers on site is invaluable, and the use of locally based firms a wise choice in terms of efficient time and construction management.

In order to strengthen the relations between the project team and to create an open shared database, an Internet blog was introduced by the contractor. The blog mapped the progress of the construction work and included comments by different team members. At the end of the project, an open day was organized where site visits and presentations helped to disseminate the knowledge gained from the use of innovative materials and technologies, although due to the disruption to the tenants, the places were limited to the press, and political and social housing organization representatives rather than the general public.

4. Characteristics of the Household's Energy Use Behaviour

In order to get an idea of the actual energy consumption after such a retrofit, the case study was monitored using data loggers and comfort surveys before and after the construction works. Data loggers were calibrated before the study, taking temperature measurements at 10 min intervals for 24 h during 7 days, before and after the retrofit.

Pre-retrofit recordings show a large standard deviation in temperatures, indicating an unstable and a rather uncomfortable indoor environment (Figure 5). As previously mentioned, in Section 3, the tenants were an unemployed couple (aged 43 and 35) and their three sons (aged 16, 11 and 4). Although the couple spent most of the time at home, they say that they turn the heating on in winter for only 5.5 h a day at 20°C. Figure 5, however, shows that indoor temperatures are very high, indicating that the heating must have been set above 20°C, possibly at 25°C. Thermal questionnaires, which were completed by the occupants three times a day, show that the children had less tolerance for temperature change and had the highest mean comfort temperature, and they also wore less clothes than the parents at home.

The monitoring process was repeated after the retrofit was completed in July 2010 and after the wall *U*-value had been improved from 2.10 to 0.12 W m^{-2} K^{-1}. Figure 6 shows that the indoor temperatures are more stable than before the retrofit. Temperature variation during 1 day is <5°C, and temperatures measured in different rooms are closer to each other (<3°C difference).

The installation of PVs may affect the energy use behaviour; according to Keirstead (2007), the installation of PVs and system performance monitors encouraged households to reduce their overall electricity consumption by approximately 6% and shift demand to times of peak generation, but the research of Bahaj and James (2007) suggests that there is little evidence of load matching or switching in the social housing sector, especially among the high demand users. After the retrofit and the installation of the Smart Meter and the PVs, the case study family was asked whether they would shift their electricity demand during peak hours: they replied that they probably would if they noticed any

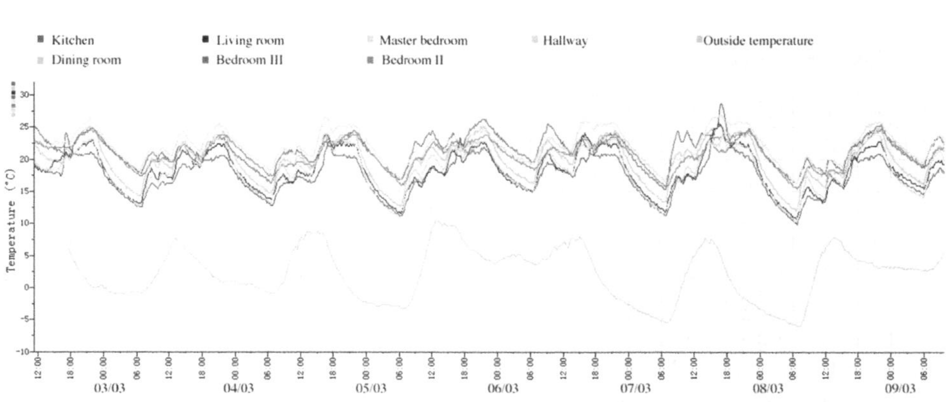

Figure 5. Pre-retrofit temperature recordings in the case study

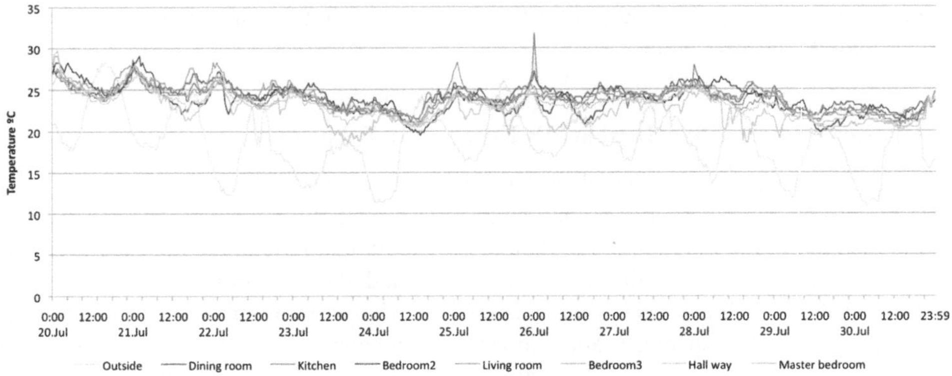

Figure 6. Post-retrofit temperature recordings in the case study

change in their electricity bills. Since there were no instructions on the PV use taking place during the interview, the tenants were not fully aware of the savings due to the load shift, nor were they able to understand the meter readings. The monitoring will continue for 2 years.

In order to better understand the characteristics of the case study household compared with other similar properties in the area, a short behavioural survey was carried out in 13 identical Council-owned properties in Trumpington. Within 13 households, 39 individuals responded to the questionnaire on heating, the use of appliances, occupancy time, level of clothing and heating habits. According to the survey, the temperature setting of the households exhibits a large variety from 16°C to 25°C, the case study being in the higher end of this scale. The number of heating hours also shows a wide range of behaviours, ranging from 5 h (as in the case study) up to 24 h per day. Due to the small sample size and time limits, it is not possible to generalize behavioural patterns, but the comfort perception of the participants does seem to differ remarkably. Further research is needed to determine the magnitude of behavioural rebound caused by the building upgrade and improved thermal performance. Moreover, a thorough study involving a larger sample should aim at a generalization of behavioural patterns and comfort perception as well as their causes.

5. Policy Strategies

After the retrofit, a Smart Meter that has a touch screen featuring an interactive graphic display connected to all the electrical appliances, the lighting circuit, the heating system and the PVs was placed. It provides an in-depth view of energy flow in the household, giving details on where most of the energy is being consumed and where wastage is occurring. At present, a PPM client who tops up weekly does not get any feedback on their energy consumption, any energy bills or annual meter readings. Driven by the 2006 EU Energy Services Directive, the Department for Energy and Climate Change has announced that gas and electricity Smart Meters will be rolled out by energy suppliers to every home in Britain by the end of 2020.

The potential impact of monthly feedback on energy use patterns is usually estimated to be 5–10% (Darby, 2006). During the first visit after the retrofit, the meter was not yet fully installed and operated in the default mode, and in the second visit, the tenants had deactivated it because of the interference with the Internet connection. The tenants expressed no interest in the meter. The fact that they had not personally invested in that technology could be one reason for the lack of motivation. Contrary to the Smart Meter, however, the tenants showed a great interest in the new "A++"-rated kitchen appliances, read the instructions and operated them successfully right after the renovation. The tenants felt more confident with white goods they were already familiar with than with the newly installed ones. Technically, Smart Meters are advanced, but the user interface of the systems leaves a lot to be desired; at this moment, the meters tend to give raw, unprocessed data in figures. As it is, many households are not even interested to programme their thermostats, and it is estimated that up to 50% of the households in the UK keep the factory settings (Pollitt, 2010). Compared with processing data from a Smart Meter, opening a window is adaptive behaviour that is intuitive and does not need data feedback—the change is expressed in the improved comfort. Previous studies have found that higher levels of education are related to less hours of using heating at the highest temperature, for example, in the Netherlands (Guerra Santín, 2010). The government policy of installing Smart Meters needs to be supported by education, especially in the social housing sector. There is a great potential in behaviour: in Britain, for every degree that the thermostat is turned down, the heat loss decreases by about 10%, and turning the thermostat down from 20°C to 15°C would nearly halve the heat loss (MacKay, 2009).

Due to high costs (nearly £6000 per installed kWp), PVs, as installed in the case study, remain unaffordable in most social housing retrofits. In order to address this barrier, the FIT for electricity came into force in the UK in April 2010. The policy requires energy suppliers to pay householders who generate their own electricity from renewable or low carbon sources. For solar power, the FIT varies from 29.3 to 41.3 p kWh^{-1}, depending on the type and size of the PV system and the date of installation, with a period that can be 25 years.

The tariffs should cover the capital cost and earn an annual return of up to 8%, in addition to the energy bill savings. Judging from the PV meter readings after the retrofit, it was estimated that the monthly economic benefit from the PVs in July will be approximately £72, with the generation tariff (FIT) of 41.3 p kWh^{-1}. This will fall by approximately 81% during the winter months. Based on the simulation, the households should be able to get up to half of their electricity from PVs and thus for free, resulting at an annual saving of around £775. In the case study, the benefit of any extra electricity sold to the grid will go to the owner, the Cambridge City Council. If a household is categorized as fuel-poor, a tenant could gain the profit from selling any surplus electricity to the grid. Such an economic incentive could trigger their interest in the system efficiency, promote load matching and, in the long term (if combined with appropriate education and feedback), lead to a behavioural change—and from a broader perspective contribute to social cohesion in deprived areas. In this case, however, there is a need to consider capital funding of PV array.

Like the case study household, in total, 12% of all energy customers in the UK (mostly low income or in debt) households are on PPMs and pay higher energy prices than if they were on a contract. Data from Brutscher (2010) show that PPM households like to make frequent and small top-ups regardless of the increased energy tariffs or income level, possibly due to liquidity constraints and loss aversion. In the case study area, families topped up their meter with around £20 weekly during the heating season. There was no indication on speculation on increasing energy prices or interest rates. Therefore, the theory of Uitdenbogerd *et al.* (2007) that energy-relevant household behaviour seems to be of habitual character rather than rational seems to be correct at least in the context of this case study.

6. Conclusions

This paper investigated the implementation of an energy-efficient retrofit in social housing and the related energy use behaviour in a case study in the UK. The conclusions are drawn based on the on-going "Retrofit for the Future" project funded by the TSB.

The energy-efficient retrofit, which improved the building to an "A" level energy performance rating, provided the occupants with a more stable and satisfactory indoor thermal environment and a lower energy bill. This was achieved by increasing thermal insulation and air-tightness levels (Section 3) to minimize the heat loss and cold bridges from the building envelope, supported by the installation of an energy-efficient boiler, PVs, LED lighting and "A++"-rated white goods. The other similar BISF houses owned by the City Council in the area will follow the example of the case study house and be upgraded in a similar, although less complicated, strategy based on the experience gained (i.e. leaving out the replacement of the existing double-glazed windows and the kitchen refurbishment to save both time and costs). The additional insulation and the LED lighting seem to be more robust, cost- and time-effective measures and do not "require particularly green tenants". Despite the FIT and the Renewable Heat Incentive, PVs and solar thermal systems remain expensive technologies for social housing and require special briefing for occupants.

Despite their original motivation, the tenants found the retrofit process difficult and disruptive. The long duration of the retrofit was considered as a barrier also by the design team, who commented that in future projects, the length should be decreased, while the occupants should be clearly informed in advance how the process is going to affect them. Further training for the planning authorities and more accurately priced costs by the contractor were two other important points of improvement.

The results support the research of Guerra Santín (2010) that the quality of building construction plays only a limited role in determining an actual energy performance in domestic buildings. The survey of households living in a social housing identical to the case study exhibited a large variety of heating patterns (Section 4). This was due to different lifestyles, occupational patterns and comfort expectations and independent of the buildings' physical characteristics. For example, families with the highest energy demand in the survey heat up their house more than three times longer than the family with the lowest energy demand. Households can use three or more times as much energy for heating as their neighbour, even if they lived in identical homes (Gram-Hanssen, 2010). Social housing may be especially heating-intensive since inhabitants, if unemployed as in the case study, tend to spend a lot of time at home. The unstable energy consumption habits in similar houses suggest two things. First, sufficient deviation in comfort temperatures should be allowed after a retrofit, more than what the standards of conventional comfort theory may indicate. People tend to be more tolerant in their comfort zones if they know that they can control the temperature and ventilation, so retrofit strategies should offer adaptive opportunities and not be over-engineered. Secondly, it is very hard to estimate standard energy consumption for even identical buildings in simulations or policies related to retrofit. Lifestyle changes have been found to be more effective than increasing the thickness of thermal insulation (9% reduction) as a means to save energy (Shimoda *et al.*, 2003). When a building's thermal properties are improved, tenants may opt for different indoor thermal conditions (e.g. higher indoor temperatures), which can cause a rebound effect in energy consumption (Guerra Santín, 2010). This calls for further research, and finding out long-term behavioural patterns is the focus in the further monitoring of the case study that will continue until 2012.

The case study demonstrated the challenge of reaching the savings in the existing housing where nearly every house, budget, occupants and the design team are different. According to Boardman (2007), 80% cut in CO_2 emissions by 2050 requires a radical new approach.

General barriers to energy efficiency that remain after the retrofit are high thermal comfort standards, high occupancy and an increasing number of (entertainment) electronic appliances. It should be noted that these barriers are not technical or mechanical but cultural and may offset energy savings from technical measures, unless energy and planning policies recognize the importance of energy use behaviour. While there is a general agreement that occupant behaviour is a major determinant of energy use in buildings, assessment of the potential reduction of CO_2 emissions with non-technological options has been limited and the potential leverage of policies over these is not well understood (Levine *et al.*, 2007). In the case study, despite being a fuel-poor household and PPM clients who end up paying more for their energy, the tenants' energy use is characterized by high indoor temperatures (reaching 25° during the heating season) and a significant number of entertainment appliances, every member of the family has a TV set—children even have their own mini-fridge. Despite economic constraints, there seems to be an acceptance of energy wasting behaviour that is passed on to the next generation who has even higher expectancy of comfort. Policy instruments such as FIT and Smart Meters are based on the rational choice models that assume that people make rational decisions, but in practice, there seems to be irrational economic behaviour. Furthermore, there was a lack of interest in feedback technology; the tenants became instantly confident with the new electrical appliances, but did not interact with the technology unfamiliar to them, such as PVs and the Smart Meter (Section 5). This indicates the need for support in the use of new equipment without precedents if the government policy of Smart Meters is to be successful.

References

Bahaj, A. S. & James, P. A. B. (2007) Urban energy generation: The added value of photovoltaic's in social housing, *Renewable and Sustainable Energy Reviews*, 11(9), pp. 2121–2136.

Bergman, N., Hawkes, A., Brett, D. J. L., Baker, P., Barton, J., Blanchard, R., Brandon, N. P., Infield, D., Jardine, C., Kelly, N., Leach, M., Matian, M., Peacock, A. D., Staffell, I., Sudtharalingam, S. & Woodman, B. (2009) UK microgeneration, Part I: Policy and behavioural aspects, *Proceedings of the Institution of Civil Engineering: Energy*, 162(1), pp. 23–36.

Boardman, B (2007) *Home Truths: A Low Carbon Strategy to Reduce UK Housing Emissions by 80% by 2050* (Oxford: University of Oxford's Environmental Change Institute).

Boardman, B., Killip, G., Darby, S., Sinden, G., Jardine, C. N., Hinnells, M. & Palmer, J. (2005) *40% House Report* (Oxford: Environmental Change Institute (ECI).

BRE on Behalf of DECC (2009) The Government's Standard Assessment Procedure for Energy Rating of Dwellings 2009. Available at http://www.bre.co.uk/filelibrary/SAP/2009/SAP-2009_9-90.pdf (accessed 2 December 2011).

Brutscher, P.-B. (2010) Pay-and-pay-and-pay-as-you-go: An exploratory study into pre-pay metering, Presentation at Cambridge University Energy and Environment Seminar, 5 May.

Communities and Local Government (2010) *Communities and Local Government Adaptation Plan* (London: Department for Communities and Local Government).

Darby, S. (2006) *The Effectiveness of Feedback on Energy Consumption, A Review for DEFRA of the Literature on Metering, Billing and Direct Displays* (Oxford: Environmental Change Institute (ECI).

Egmond, C. & Bruel, R. (2007) *Nothing is as Practical as a Good Theory, Analysis of Theories and a Tool for Developing Interventions to Influence Energy-Related Behaviour* (Utrecht: SenterNovem).

Galvin, R. (2010) Thermal upgrades of existing homes in Germany: The building codes, subsidies, and economic efficiency, *Energy and Buildings*, 42(6), pp. 834–844.

Gram-Hanssen, K. (2010) Residential heat comfort practices: Understanding users, *Building Research and Information*, 38(2), pp. 175–186.

Guerra Santín, O. (2010) *Actual Energy Consumption in Dwellings. The Effect of Energy Performance Regulations and Occupant Behaviour* (Amsterdam: IOS Press BV).

Haas, R. & Biermayr, P. (2000) The rebound effect for space heating: Empirical evidence from Austria, *Energy Policy*, 28(6–7), pp. 403–410.

HM Government (2010) *Warm Homes, Greener Homes: A Strategy for Household Energy Management* (London: Department of Energy and Climate Change/Communities and Local Government).

IEA (International Energy Agency) (2008) *Promoting Energy Efficiency Investments, Case Studies in the Residential Sector* (Paris: IEA/OECD/AFD).

Keirstead, J. (2007) Behavioural responses to photovoltaic systems in the UK domestic sector, *Energy Policy*, 35(8), pp. 4128–4141.

Levine, M. D., ürge-Vorsatz, P., Blok, L., Geng, D., Harvey, S., Lang, G., Levermore, A., Mongameli Mehlwana, S., Mirasgedis, A., Novikova, J., Rilling, H. & Yoshino, H. (2007) Residential and commercial buildings, in: B. Metz, O. R. Davidson, P. R. Bosch, R. Dave & L. A. Meyer (Eds) *Climate Change 2007: Mitigation, Contribution of Working Group III to the Fourth Assessment Report of the Intergovernmental Panel on Climate Change*, pp. 418–431 (Cambridge: Cambridge University Press).

Liao, H. C. & Chang, T. F. (2002) Space-heating and water-heating energy demands of the aged in the US, *Energy Economics*, 24(3), pp. 267–284.

MacKay, D. (2009) *Sustainable Energy Without the Hot Air* (Cambridge: UIT Cambridge).

Meijer, F., Itard, L. & Sunikka-Blank, M. (2009) Comparing European residential building stocks: Performance, renovation and policy opportunities, *Building Research and Information*, 37(5), pp. 533–551.

Pollitt, M. (2010) The economics of energy demand, Presentation in EPRG Spring Research Seminar, Cambridge, 13–14 May.

Shimoda, Y., Fujii, T., Morikawa, T. & Mizuno, M. (2003) Development of residential energy end-use simulation model at a city scale, in: *Proceedings of the Eighth International IBPSA Conference*, Eindhoven, the Netherlands, 11–14 August 2003.

Sunikka, M. (2006) *Policies for Improving Energy Efficiency of the European Housing Stock* (Amsterdam: IOS Press).

Tambach, M., Hasselaar, E. & Itard, L. (2010) Assessment of current Dutch energy transition policy instruments for the existing housing stock, *Energy Policy*, 38(2), pp. 981–996.

Uitdenbogerd, D., Egmond, C., Jonkers, R. & Kok, G. (2007) Energy-related intervention success factors: A literature review, in: *Proceedings of the ECEEE 2007 Summer Study: Saving Energy—Just do it!*, La Colle sur Loupe, France, 4–9 June 2007.

Mediating Low-Carbon Urban Transitions? Forms of Organization, Knowledge and Action

MIKE HODSON & SIMON MARVIN

SURF, University of Salford, Salford, UK

ABSTRACT *Increasingly at the scale of cities, strategies and plans to respond to the challenges of climate change and constrained resources are being developed. A range of climate change plans, low-carbon strategies, peak oil preparations and so on have been developed, often with ambitious aspirations. At the same time, new and reconstituted "intermediary" organizational forms are working between the priorities of these plans and the contexts of their "application". This is the movement between the "what" of the plans, strategies and preparations and the priorities they embody and the "how" of attempts at their accomplishment. Drawing on research in Greater Manchester, in this paper we examine the organizational contexts constituted for such a purpose and ask fundamental questions about whose priorities are being advocated, where and how this is organized and what the implications of this are for forms of urban transition.*

1. Introduction

This paper contributes to an emerging but increasingly vibrant debate about the role of cities in transitions to low-carbon energy futures (Bulkeley *et al.*, 2010; Hodson & Marvin, 2010a, 2010b). In doing so, it starts from the view that there are many ways that energy infrastructures, that support the social and economic life of the city and that produce particular ecological consequences, can be shaped and that potentially different coalitions of social interest can claim to speak on behalf of the city. Consequently, this paper's focus is on the intersection of two sets of issues. The first issue is the relationship between city and energy system in a period of conflicting economic, ecological and political pressures and the search for a low-carbon future and the second is the social interests, institutions and actors, who seek to shape such a relationship and the ways in which they are organized to act in doing so. In that sense, our concern in the paper is with: assessing the "intermediary" contexts constituted for the purpose of low-carbon urban transitions;

understanding the politics of whose priorities are dominant in these contexts; and what the implications of this are for urban transitions.

Cities and regions are integrated into wider energy systems as key sites of consumption in complex socio-technical networks and they are implicated in hierarchies of multi-level governance. There are clearly different scales of action—national, regional, city-regional, publics, etc.—involved in coordinating energy systems and the transition to low-carbon futures and there are likely to be different strategic priorities at these multiple scales of action. This emphasizes the importance of understanding the relationships between these scales and how and why they are coordinated, in tension or even disconnected. It involves examining the relationships between national priorities and plans and how these are interpreted and responded to at an urban level and also the extent to which cities' strategies and national plans and priorities can be better integrated. Consequently, understanding whose priorities it is that shapes the activities that aspire to constitute a particular low-carbon urban energy transition is an important issue.

Analytically the focus of this paper is whether new organizations are created to enact a low-carbon urban transition or whether existing organizational contexts are adapted to bring together (or exclude) a range of national, local, regulatory, utility and other relevant social interests. This means understanding the type of low-carbon energy future—and its relationship to the city—that is envisaged but also the social interests who are favourably positioned and disposed towards such a future and the processes of participation that result. This is particularly important in a wider context of shifting landscape conditions and whether the transition envisaged represents something fundamentally different from the existing energy system. Our concern is with how these multiple social interests are organized and how their different knowledges are organized in the capability to act.

The paper does this through a case study of the early stages of attempts to undertake a low-carbon energy transition in Greater Manchester. Greater Manchester has sought in recent decades to re-position itself from an industrial city to a post-industrial entrepreneurial city where new formal metropolitan governance arrangements that have been absent since 1986 have since the late 2000s been under development. This changing governance context has a significant emphasis on developing low-carbon energy capabilities alongside a range of other, primarily economic, priorities. This means that the power relationships between national government and "local" social interests are frequently negotiated and played out through these different priorities. This focus on the dynamics of transition in Greater Manchester provides a context to understand the "type" of transition envisaged, the social interests behind such a transition and what the likely consequences of it are. Exploring these issues in relation to a changing governance context provides the potential to offer lessons for other low-carbon urban transitions in different contexts.

The remainder of the paper is divided into five parts. First, we review the literature on socio-technical transitions and identify the implications for place-based transitions. Second, we use documentary analysis of national policy documents to identity national priorities around energy and climate change and the envisaged role of place-based transitions. Third, we use documentary analysis, observational material and an interview programme with city-region-scale actors to analyse the forms of experimentation taking place in the emergent Greater Manchester city-region as part of attempts to forge a low-carbon city-region. Fourth, we identify the key issues from the Greater Manchester case study for understanding whether a urban low-carbon transition has—or is—been enacted. Finally, we finish by offering key conclusions for low-carbon transitions and identify issues for future research.

2. Place-Based Low-Carbon Transitions and Organizing Action

Transitions analyses, both historical and prospective, have been put to work across a range of substantive technological, systemic and sectoral areas (Geels, 2004; Voß *et al.*, 2006; Smith *et al.*, 2010). Transitions approaches, through their three-fold concepts of landscape, regime and niche and their interrelated mobilization as a multi-level perspective (MLP), have also allowed researchers to examine the role of experiments and niches in relation to the existing socio-technical regimes and the dynamics between regime lock-in and niches as the basis for regime transformation. The potential of the MLP has been highlighted in "its use as a flexible heuristic" (Smith *et al.*, 2010, p. 436) and through the interest shown in it by a wide range of researchers and policy interests.

A central issue for us is to what extent the MLP can be utilized to understand the development of place-based low-carbon activities. In particular, analysing the extent to which transitions can be undertaken at the level of the city is important because of the complex pressures and tensions for urban economic growth, national targets for reducing subnational carbon emissions and developing effective responses to the threats of climate change (Hodson & Marvin, 2010b). Urban authorities and wider coalitions, in particular in world cities, have sought to (partially) reconfigure energy systems at the level of the city; they have done this and are doing this to try and assert control over the organization and functioning of energy systems to build greater security of supply for cities and the social interests who benefit from urban growth and also to meet carbon reduction targets. The difficulty of this, in a UK context, is that the promotion of a competitive city-based economic growth takes place in a broader context where national government engages with cities in national space differentially but also where energy regimes or systems are not organized at a city-regional scale.

The relationship between national government and urban actors is underdeveloped in transitions approaches (Coenen *et al.*, forthcoming). The MLP has primarily been concerned with national-level transitions. There are important issues raised by the ways in which national governments view the role of cities and communities in undertaking transitions. Re-organizing energy systems or effecting a transition at the city scale, therefore, necessitates a transition from a largely regionally nationally organized energy regime in a wider national system but also requires the constitution of a city-scale regime with variable levels of discretion afforded by national government. In this sense, when we talk about an urban low-carbon transition we are referring to a re-scaling of the energy regime, in ways which transform the city as well as the energy regime and that also requires the development of—and the "intermediary" organization of—the capacity to act in undertaking such a transition.

Yet a number of recent contributions have highlighted that locating the role of the city—theoretically, conceptually and empirically—in low-carbon transitions is extremely difficult (Bulkeley *et al.*, 2010; Coutard & Rutherford, 2010; Hodson & Marvin, 2010a; Späth & Rohracher, 2010). The MLP says little explicitly about cities, who, what and where the city is, and their roles within transitions.[1] Locating the city in low-carbon transitions, given its relative neglect in the MLP, is likely to be the subject of politics and struggle among researchers seeking to understand the role of cities in transitions and also between policy-makers and other interests engaged in the practice of transitions. The MLP provides us with a foil—a "flexible heuristic" (Smith *et al.*, 2010)—through which to think through the roles of the city in low-carbon transitions.

Given the politics and struggle between differentially positioned institutions and social interests involved in transition activity, whether transitions can be managed in the sense of purposively steered or not is a difficult issue (see Shove & Walker, 2007). This should not be treated as a simple yes or no question. There are significant positions in between, including the view that transitions can be the consequence of purposive intent, unintended consequences and subsequent adjustments. This acknowledges the teleological leanings of accounts that view the management of transitions in time-bounded ways rather than viewing time horizons as looser reference points and as part of ongoing and unfolding processes of transition.

Whether the city is viewed as an actor in its own right, a niche for experimentation to think about new ways of organizing relationships between energy producers, consumers and flows through the city, a regime of existing relationships between energy producers, consumers and flows through the city or any other conceptualization has important consequences for how a series of contexts and relationships within the city are understood. The MLP can further assist us here in thinking about what broader changes in political economy are contributing to. How, for example, landscape economic and ecological conditions shape attempts to reconfigure energy systems at a city scale. The ways in which reconfiguring energy systems is as a means of securing the low-carbon resource flows to literally fuel new economic accumulation strategies. But also the ways in which the development of new accumulation strategies, through low-carbon transition, then start to feedback into changes in landscape conditions through re-enforcing the pre-dominance of the ideology of competition between places as underpinning economic growth.

Yet the framing of low-carbon transitions can be manifold and often the basis for different framings is the participative constitution of "visions" by transition actors (Hodson *et al.*, 2010). The act of representing low-carbon futures through the mutual future of city and infrastructure is underpinned by political efforts to bound time–space through socio-technical transition in the service of particular social interests. The concept of "vision" in the MLP is important as "the articulation of visions and expectations to provide an orientation towards the future and give direction to learning processes" (Geels, 2005, p. 366). Yet, particularly given the range of regime interests—many of whom have incentives to defend the status quo—urban decision makers and potential new interests with motivations for a new or reconfigured regime, it is important to recognize the struggles and negotiations that inform the production of both dominant and competing visions. Visions, in this understanding, are re-cast as the symbolic representations of the future relationships between city and regime that are produced through these relational struggles to define and categorize.

Seeing a vision as representational space (Massey, 2005) means that complex and deeply political relationships between an energy regime and the city can be articulated and framed in highly exclusionary ways that are presented as fixed, closed, static and discrete. It also excludes a multiplicity of co-existing narratives and potential narratives. Seeing space as actively, relationally and relatively produced (Massey, 2005; Harvey, 2006) offers the possibility to understand attempts to define and categorize the relationship between a future energy regime and the city. It offers this not only in relation to existing and absolute representations of regime and city but also through the ways in which the multi-level governance field of interrelationships of institutions and social interests seek to re-constitute a mutual identity for the regime and the city—this we shall illustrate in Section 4.

The field of such social interests (Bourdieu, 1993) is not an equitable one and highly particular coalitions of social interests, in relation to particular places, are often able to mobilize financial, relational and knowledge resources through which they produce the symbolic "visions" of what the low-carbon future of the city should be (Hodson *et al.*, 2010). Whether these visions are genuinely participatory and inclusive and link to processes of mobilizing effective capability to enact a transition or whether they are largely representations of the future relationship of city and energy regime that are produced by narrowly constituted social interests are, again, parameters which contain a series of potential intermediate positions. Underpinning this are different "types" of "intermediary" organizational contexts and cultures that mediate and through which social interests and a range of resources coalesce (Hodson & Marvin, 2010a).

Low-carbon urban transitions are then about competing views of the role of the city, the type of transition that is deemed to be required, the politics of participating in producing a "vision" of the future, how to translate that vision, and, therefore, the variability of the consequences of a transition. What we seek to do in the remainder of this paper is to focus on Greater Manchester to understand the multiple organization of "transition" activities within the city; to understand the politics of the role of the city, dominant visions of its low-carbon future, messy attempts to translate such a future from national political priorities, to "shared" city-regional priorities and subsequently to begin the process of their material manifestation; and the consequences of this. In the next section, we review the different national policy priorities and the ways in which they provide a context of enablement and constraint in Greater Manchester.

3. National Low-Carbon Policy Priorities and the Role of Place

Understanding recent national low-carbon/place policy priorities in the UK is far from straightforward. To take the area of energy policy, for example, until the Department of Energy and Climate Change (DECC) was created in 2008 there had not been a solely designated Department of Energy in the UK since 1992. This meant that historically government priorities around energy were formulated in a multiplicity of departments which had a range of issues as their core brief—trade and industry, environment, food and rural affairs, etc. The consequence of this is that current UK priorities around energy need to be pieced together from a variety of different departmental positions. In relation to the current and future shape of pressures on the UK's energy systems and the priorities that are formulated by the UK government, four departments are of particular importance: DECC; the Department for Business, Innovation and Skills (BIS); the Department for Environment, Food and Rural Affairs (DEFRA) and the Treasury (HMT). Additionally, in relation to territorial aspects of the UK government and relationships between national government and cities and regions, departmental priorities are particularly associated with the Department for Communities and Local Government (DCLG) and HMT. Experimentation with the governance of territorial and energy priorities are mediated through five dominant strategies (Table 1).

Table 1. UK energy priorities and their relationship to territory

Plan	Priorities	Promoter	Spatial concept	Exemplification
Climate Change Act; Energy Act 2008	Binding, long-term statutory greenhouse gas emissions reductions—80% 2050; 34% 2020—carbon targeting and budgeting	DECC/ DEFRA	Potential for budgeting, targets and the cascading of these down through various territorial tiers	Place-based low-carbon budgets
UK Low Carbon Transition Plan (from July 2009)	Long-term transition plan to a UK low-carbon future	DECC	Multiple views of places both implicitly and explicitly	Community pilots; competitions for towns and cities
Low Carbon Industrial Strategy (from July 2009)	Low-carbon industrial interventionism	BIS/DECC	LCEAs	Northeast England; southwest England; M4 Corridor; Greater Manchester
Statutory City Region Pilots (from April 2009)	Design and piloting of city-region governance structures for sustained economic growth	Treasury/ DCLG	City regions	Manchester; Leeds
Re-designing Sub-national Governance (from Spring 2010)	Stripping out sub-national institutions, setting up LEPs, to promote new forms of sub-national economic activity and competition	BIS/ DCLG/ Treasury	LEPs	LEPs—abolishing intermediate governing architecture and construction of "market" framework

3.1 *Setting Parameters for Urban Responses Through Targets, Plans and City-Regions*

The significance of the 2008 UK Climate Change Act was in its positioning of the UK as the first country in the world to have a legally binding framework for cutting carbon emissions. It does that through setting legally binding targets, creating powers to address those targets, providing the institutional framework to underpin the achievement of these targets and to do so in ways which not only sets out the UK's response to climate change but which is accountable to the UK Parliament.

The binding targets for the UK mean that regardless of whether and on what scale there is international action on climate change the UK must act. Among the key priorities in the Act is the setting of legally binding greenhouse gas emissions reduction targets of at least 80% by 2050 with an interim reduction in emissions of at least 34%, from a 1990 baseline, by 2020. This is to be achieved through 5-year carbon budgeting systems. These developments create new pressures relating to climate change and carbon regulation (While, 2008). In addition to statutory carbon reduction targets cascaded down from international agreements (Bulkeley and Betsill, 2003) those developed by national government place renewed emphasis on to sub-national territorial units, and will then place a premium on the ability of states and territories to better manage energy consumption and accelerate the development of low-carbon energy transitions.

In July 2009, the UK government published its Low Carbon Transition Plan (LCTP) which detailed broadly how the UK would meet the 2020 and 2050 emissions reduction commitments set out in the Climate Change Act. The Plan is underlain by five stated principles: first, to protect the UK public from the immediate risks of climate change; second, to anticipate how the consequences of climate change are prepared for, particularly in relation to infrastructure and housing; third, that climate change requires a new international agreement on global emissions reduction; fourth, that the UK can play its part by developing a low-carbon country to meet targets set out in the Climate Change Agency (CCA) and address vulnerabilities and promote economic opportunities; fifth, that addressing climate change requires widespread participation from communities, businesses, individuals and so on (BIS/DECC, 2009). LCTP is not only a transition route map to 2020 for the UK, but also operates in prioritizing the carbon savings expected across different sectors.[2]

The Low Carbon Industrial Strategy was launched jointly by BIS and DECC in July 2009. It aims to position British businesses to secure the economic and job creation opportunities of a low-carbon transition and, in doing so, to minimize the economic costs of inaction. The strategy details a range of potential low-carbon sectors and technological areas—wind, wave, tidal, low-carbon vehicles, carbon capture and storage, etc.—and also a more strategic approach to the development of low-carbon economic activity and technologies across the regions of the UK, particularly through designating low-carbon economic areas (LCEAs).

A less directly "interventionist" national government role in sub-national activities was outlined in 2008 in the UK government's support for the creation of two city-regions in 2009 and, in doing so, the development of new metropolitan governance structures. The broad parameters within the city-regions were to operate as statutory forms of sub-regional cooperation between local authorities with the aim of them being significant contributors to sustainable forms of economic growth. Low-carbon economic activities were also worked into these proposals as city-regions took a more active role in shaping low-carbon transition in their own contexts.

3.2 *From Targets and Plans to Practice: New Industrial Interventionism or Re-designing Sub-national Governance?*

The priorities of the Climate Change Act and its emphasis on emissions reduction are broadly supported across UK political parties. There have, similarly, been few dissenting political voices in relation to the LCTP. The principal political tension is in the process of how the strategic priorities will be achieved—what are the mediating frameworks and institutions, what economic, social and knowledge resources are allocated to them? While there appears, superficially at least to be a broad consensus around the policy priorities—the achievement of large emissions reduction—the central controversy is around the mode of governing and the ways in which "intermediary" activity is organized.

The tension is between, first, new forms of state industrial interventionism in regions, city-regions and pan-regions and, second, in constructing new forms of national-sub-national governance fixes that sees the state less in the direct role of industrial intervener and more re-cast as a "facilitator" for city-regions and local economic partnerships (LEPs) to create the conditions for market-based and private-sector led activity. This struggle was inherent within the Labour government, which governed until 2010, and cut across its different strategies. The subsequent coalition government, from May 2010, is comprehensively re-designing sub-national governance by actively seeking to abolish and re-design institutional mediators between its central departments and places to create the conditions to compete for limited resources and create private and entrepreneurial responses that will "emerge" and develop place-based low-carbon activities. In short, the existing dominant mediators of national-sub-national relations—Regional Development Agencies—are being abolished and replaced by LEPs. At the same time, a much less well-resourced Regional Investment Fund—c£1 billion—will intensify competition between places for national resources and support.

The national policies and priorities that we have reviewed encompass a range of economic, environmental, technological and territorial issues. The ways in which different priorities coalesce within the context of a particular city-region are unclear but are likely to involve the negotiation of these different national priorities with fledgling city-regional priorities. The mediation of these city-regional priorities, their organization as capability to act and the ways in which they interact with national priorities is the issue to which we now turn in relation to a case study of Greater Manchester.

4. Mediating a Low-Carbon Energy Transition in Greater Manchester?

The national priorities and strategies outlined set the parameters within which an understanding of the responses at the city scale need to be framed. In this section, we address how city-regional capability to act is developed in Greater Manchester,[3] what a low-carbon future looks like and how it is made. The development of a response to climate change is deeply intertwined with new city-regional governance structures more generally. Such responses will be in part about the implementation of national policy and targets and part concerned with territorially specific objectives. In this respect we examine how the boundary between the disciplinary role of national government and a more discretionary city-regional agenda is negotiated.

We do this in a context where the energy regime is organized in ways that make city-regional control of such a regime extremely challenging. By regimes we are talking about the socio-technical energy regime, which are "the semi-coherent set of rules carried by different social groups. By providing orientation and co-ordination to the activities of relevant actor groups, socio-technical (ST)-regimes account for the 'stability' of ST-configurations". These configurations will contain actor-networks of producers, users, policy-makers and public authorities, suppliers of materials and components, financial capital and research and development. "This stability is of a dynamic kind, meaning that innovation still occurs but is of an incremental nature" (Geels, 2002, p. 1260).

To illustrate this, for example, the electricity regime that supplies the citizens and businesses of Greater Manchester is organized through a private regional utilities, national transmission and power producers. In the northwest of England region electricity is generated primarily through a combination of four large power plants (one coal, two gas and one nuclear), but also through medium-sized combined heat and power (CHP) plants and smaller embedded generation of renewables and small-scale CHP. There is a high-voltage electricity supply network and a low-voltage electricity supply system, each operated by different commercial interests. Within this system there is limited spare capacity in the regional high voltage transmission lines which are part of a national network. Significantly, the network in the northwest mediates the distribution of electricity from Scotland which has excess capacity to the south of England where demand far outstrips electricity production in the region. The consequence of this is that the transmission network in the northwest of England is shaped by interdependencies between the north and south of the UK. Energy demand and consumption in the northwest can be understood not only with respect to domestic consumption but also through the large manufacturing base around chemicals, paper and other energy-intensive industries. There is recognition within many of these large energy-intensive industries that energy use in relation to output needs to be made more efficient and there have been efforts to this end in response to growing costs of energy and fuel but also in response to national government energy efficiency programmes.

One can see from this brief example of the regional organization of electricity that although energy may be consumed by city-regional businesses, citizens and organizations the constitution and organization of the energy regime involve many actors and issues. This includes producers, feedstocks, supply networks, multiple technologies, regulators, housing and commercial buildings, national government and consumption patterns. But also intersecting with the regime are multiple transport systems and industries that operate within and beyond the city-region, relationships with other regions and so on. This illustrates that the constitution of the regime on which the city-region is reliant is not coterminous with Greater Manchester.

4.1 *The "Vision": Climate Change as a Low-Carbon Economic Opportunity for Greater Manchester and Meeting National Targets*

The Association of Greater Manchester Authorities (AGMA) since 1986 has worked to coordinate the actions of the 10 Greater Manchester local authorities on cross-boundary strategic areas—such as transport and waste—that are seen as most effectively organized at a metropolitan scale. In January 2008, the Executive of AGMA agreed in principle to support the establishment of a CCA across the city region. The development of the CCA was the product of a confluence of activities that sought to coordinate local actions and "sustainable" economic strategy at a city-regional scale. A significant step in this was the 2008 "Mini-Stern" for Manchester commissioned by AGMA and undertaken by the consultants Deloitte. It aimed to produce a re-scaled version of the Stern Report that was written for the UK Treasury in 2006 by the economist Nicholas Stern, reviewing the economics of climate change.

The Mini-Stern sought to calculate the cost of climate change for the Manchester city-region and possible strategic responses in the short-to-medium term particularly given international pressures and national legislation on climate change. The report was framed relatively narrowly in that it examined the economic costs of inaction on climate change and the economic opportunities of early action. On the basis of analysis conducted using a methodology developed by Deloitte, it calculated that inaction on climate change would potentially cost Greater Manchester £21 billion over a 12-year period and £72 billion at the northwest regional level. It prescribes a necessity for action at the Greater Manchester level that prioritizes distinctiveness, early movement, technology-led responses, business and investment, business support and attempts to attract inward investment. The focus on energy in the review is oriented towards re-shaping city-regional energy flows in line with meeting renewable energy and other national targets.

Consequently, it argues for a strategic approach to energy production, consumption and the reconfiguration of Greater Manchester's energy system through energy efficiencies and new technologies to reduce economic cost and meet carbon reduction targets. In this respect, to achieve national targets but also to meet the city-region's energy "needs" is linked to achieving greater control over energy production and consumption relationships through coordination of urban regeneration activities and latent innovation capacity in Greater Manchester. Mini-Stern also raises the importance of eco-innovation and creating new markets for services and technologies. In doing this, the role of the public sector is promoted as an exemplar by leading through example in terms of procurement strategies and its own estate. But also this is part of an attempt to create a culture that is conducive to business investment, inward investment and the availability of support and advice. In doing this there is also an emphasis on developing and building skills and capacity—with a role for higher education institutions—and the alignment of policies. The logic of the report appears to be in extending the economic competition between places into the sphere of climate change and for Greater Manchester to seek to exploit a first mover advantage and distinctiveness in eco-economic competition. It is in line with addressing (and ignoring) many of these issues and concerns that the CCA was set up in parallel with two other energy intermediaries (Table 2).

4.2 *Making a Greater Manchester CCA: Economy and Environmental Priorities*

The CCA is closely intertwined with the development of new city-regional governance structures for Greater Manchester. It was intended that the Agency would build capacity for behavioural change work, support for businesses, planning policy and procurement and the development of critical infrastructure—this included combined heat and power and renewable energy installations. During 2008 work on developing the governance arrangements for both the Environment Commission and the CCA were led by the Environment Commission Development Team (ECDT) and in mid 2009 AGMA approved the establishment of the CCA as a Shared Administrative Structure prior to it becoming a limited company.

Although the establishment of the CCA was the remit of the Environment Commission initial development, funding for it was provided by the NW Regional Development Agency (NWDA) channelled through the city-regional economic partnership Manchester Enterprises (ME)—which was the forerunner to the Economic Commission at city-regional level. Part of this funding stream supports a Carbon Manager within the Economic Commission to support the ECDT. Importantly, we can see how the funding streams and their different strategic priorities begin to contribute to the constitution of a CCA that blurred the distinctiveness of the Environment Commission and the Economic Commission as well the environmental and economic priorities underpinning each commission. These, of course, are not necessarily in conflict—in fact the Mini-Stern would suggest they are complimentary—but how these interrelationships were understood and developed is particularly significant for the type of approach to climate change that was produced. The parameters and activities of the CCA were developed not only through these processes and relationships within AGMA, the NWDA and ME, but also through discussions with central government. The aim was that by summer 2010 the CCA would be established as a legally constituted body.

Table 2. Greater Manchester "low-carbon" "intermediaries"

Organization	Priorities	Promoter	Spatial concept	Implications
Greater Manchester CCA	Develop capacity for behavioural change work, business support, planning policy and procurement—development of critical infrastructure including CHP and renewable energy installations	City-regional (AGMA)—Environment and Economic Commissions; national government (Treasury; DCLG)	Constructing the city-region through responding to climate change and developing capacity to act	The dominance of economic agendas and pre-existing economic interests in constructing a city-regional low-carbon response
Greater Manchester LCEA	LCEA "retrofit" is a 5-year initiative—operates in public and private sectors addressing insulation, smart metering technologies and small-scale renewables—key national priorities—creating low-carbon economic opportunities	National government designation LCEA proposal developed by the GM[AQ6] Environment Commission/Economic Commission with national government departments and regional and national agencies	LCEA primarily exemplified as a zone—Oxford Road Corridor—of the urban core rather than the extended city-region	To exemplify how to address national targets—to exemplify and promote systemic change—to act an attractor for inward investment
Greater Manchester Energy Group	Provide effective strategic governance working to coordinate energy and energy-related actors and issues at a city-regional scale	Intended to include energy companies, public and private sector interests, representatives from across city-region and national government—supported by officers	Strategic overseer for energy issues across GM	Although not explicitly part of the CCA, would in the "medium term" deliver the projects and programmes it developed—the work of creating this capacity is extremely challenging

The two principal roles of the CCA were seen as the delivery of new programmes devised on behalf of the Environment Commission and to develop capacity and resources to deliver existing projects. Initially, this meant bringing existing and planned inter-authority work within the CCA through a 3-year business plan. This resulted in developing a mix of new initiatives, such as the CCA and the LCEA, but that also encompassed and re-packaged existing activities across a range of partners and scales. For example, this included Greater Manchester activities in a national network of local energy efficiency advice centres, Manchester City Council's work as part of a national Low Carbon Cities Programme and Manchester is My Planet's work as part of an EU programme.

A significant part of the rationale for a metropolitan-scale CCA is that it is an appropriate strategic scale to act on climate change because effective local capacity to act on climate change was missing and that developing such capacity requires the sharing and coordination of resources, particularly in relation to energy, water, transport and emergency planning. This attempt to develop new governance structures is not without its difficulties. In relation to the challenges of climate change and energy it will require cross-commission working as responses encapsulate spatial planning, critical infrastructure, public health and economic activity.

4.3 *A LCEA—Showcasing Technologies and Meeting National Targets*

Greater Manchester was designated in December 2009 as the UK's fourth LCEA for the Built Environment. The draft prospectus for the Greater Manchester LCEA was produced by consultants Ernst and Young under commission from AGMA. The prospectus linked together the issues of carbon reduction with the existing built environment and the need for systemic retrofitting particularly through demand-side management measures—energy efficiency measures and alternative sources of heat and power in residential, private business and public buildings—and, in doing so, creating low-carbon economic opportunities. The LCEA frequently alludes to Greater Manchester's existing assets in relation to the exemplification of retrofitting. This includes the largest photovoltaic retrofit in the UK of Manchester's Co-operative Insurance tower and the proximity of leading research and development institutions, particularly in relation to the built environment. And it is also claimed that the LCEA will contribute to saving six million tonnes of CO_2, deliver up to £650 million additional gross value added, support 34,800 jobs in total, and contribute wider economic, learning and best practice benefits to the region and to the UK generally.

The LCEA prospectus pointed out that such responses were expected to contribute to systemic and transformative change in buildings, the use of energy and reduction in carbon emissions. Yet the prospectus also pointed out that systemic change required the development of a new skills capacity and understanding of the types, range and extent of low-carbon technology providers and supply chains within the city-region. If achieved building energy performance standards would be improved, mandatory carbon reduction targets would be met in a cost-efficient way, and a market developed for commercial retrofit within Greater Manchester. The LCEA proposal was developed by the city-region's Environment Commission but was also based on inputs from the Economic and Planning and Housing Commissions and also national government departments and national and regional agencies. The aim was that by April 2010 a Joint Delivery Plan would be developed that would be shared by national government and agencies, regional and local government and agencies.

The visibility and symbolic importance of the LCEA is important as the aim of the zone is also to act an attractor for inward investment. More specifically, the LCEA "retrofit" programme is a 5-year initiative that operates in both public and private sectors addressing insulation, new smart metering technologies and the implementation of small-scale renewable technologies—all national priorities. The Oxford Road Corridor in close proximity to two of Manchester's universities will provide the location for a "low-carbon laboratory" to develop and test new technologies, which specifically includes energy supply infrastructure and energy efficiency measures and offers a "working vision" of a retrofitted low-carbon city.

In the early months of 2010, the LCEA was still at the very early stages of development. Work to realize the LCEA requires partnership building and ways of working that develop the organizational, knowledge and financial capabilities to deliver it. There is no single source of funding for the LCEA. Rather the designation requires the city-region to work in partnership with numerous national actors and agencies, regional partners, and also to seek to secure funding from a range of national, regional and supranational funding streams. In this way the city-region must constantly position the LCEA to try and attract funding whilst also meet the multiple priorities and expectations of different partners.

The sectoral designation of Greater Manchester as an LCEA in the built environment and its spatial designation in part of the urban commercial and retail core appeared somewhat at odds with the city-region's initial view of a low-carbon future being developed through a variety of activities in different spatial zones across the wider city-region. The tighter spatial view that was being promoted through the LCEA strongly reflected the priorities of national government's business department whilst the broader and enlarged view would require Greater Manchester to engage with different national priorities within the community, energy and environment departments. Although the claim is that both options are being kept open—it is the LCEA agenda where more appears to be happening through the retrofit of existing domestic and commercial buildings (Report to Environment Commission on Environment Commission Work Programme: Commissioner Roles, 2009). This view is narrower in terms of territorial scope; narrower in terms of technologies; and narrower in that it seeks, for example, to deliver—and provide a model for delivering—specific national priorities in relation to smart metering and mandatory carbon reduction targets.

4.4 *Greater Manchester Energy Group—Creating City-Regional Energy Capability?*

As part of the city-region pilot Greater Manchester was committed to establishing a multi-sectoral city-region Energy Group. This would provide effective strategic governance working to coordinate energy and energy-related actors and issues at a city-regional scale. The Group would include energy companies, public and private sector interests, representatives from across the city-region commissions and from representatives of national government. The Group would be supported by officers who would work to develop and implement actions in support of energy aspects of a low-carbon economy for Greater Manchester. The background to the Energy Group can be traced through meetings in spring of 2009 within the Environment Commission. Subsequent discussions were held between senior representatives of the Environment, Economic and Planning and

Housing Commissions and also between city-region representatives and national government departments to discuss the centrality of the role of energy in the city region pilot.

The view that was laid out in the city region pilot and its precursor documents was that economic growth aspirations in an emerging low-carbon era meant that the city region required greater security and reliability of infrastructure especially in relation to energy issues. A city-regional energy plan was proposed and ongoing work on energy planning has subsequently been integrated with the proposed work programme of the Energy Group. The Energy Group, although not explicitly part of the CCA, would in the "medium term" see the CCA deliver the projects and programmes it developed.

The proposed role of the group is as a strategic overseer for energy issues in Greater Manchester including: identifying significant energy issues for the city region; representing city-regional interests at national and international levels; and, acting as a broker between the different interests necessary to overcome the difficulties in seeking to achieve a low-carbon energy system. In doing this the Group would also act as a filter for energy bids and proposals to national and European levels, highlighting and endorsing those that originate in and focus upon low-carbon energy infrastructure for the city region. Practically, this would mean that within the context of a bundle of targets being developed at, or devolved to, the city-regional level the development of a performance framework to measure targets and model the implications of different scenarios, the development of a Framework Energy Plan for Greater Manchester and related to this, attempts to engage in infrastructure delivery mechanisms. Also related are attempts to utilize the procurement activities of the public sector, through priorities and plans, with the priorities of any subsequent city region energy plan. Furthermore, the Group will seek to align corporate climate change strategies and investment programmes with the Energy Plan and also explore the possibilities that a city region-level Investment Fund that could provide financing for specific energy projects.

Such an agenda potentially cuts across a wide range of plans, activities and social interests but it does so from an existing context where there is limited capacity and capability at a city-regional scale. This means that the activities envisaged by the Energy Group require new ways of bringing together public and private sector, utility and city-regional policy interests, different scales of policy interests, universities and business and so on—so returning full circle to the need for a CCA.

In summary then three city-regional energy intermediaries are being produced by a complicated set of relationships within Greater Manchester. These involve relationships particularly between the Environment and Economic Commissions, but also between Greater Manchester and national government, Greater Manchester and the NWDA, and at local authority levels. But the development of the CCA, LCEA and Energy Group is also caught within complex and messy combinations of these multiple scalar relationships at different times and instances. These pressures can be strongly externally driven through national government priorities, as in the case of LCEA designation, yet simultaneously are internally focused on the development of new capacity and capability through the Energy Group. These arrangements are still in a considerable state of uncertainty, as the consequences of the new government public expenditure reduction, the abolition of the Regional Development Agencies (RDAs) and establishment of LEPs.

5. Whose Low-Carbon Urban Transition?

What the case of Greater Manchester demonstrates is the richness and multi-faceted character of an attempted transition in re-scaling an energy regime to a city-regional scale within a broader national state space. National government priorities in relation to state spaces and energy in the UK in recent decades have promoted competition between places and the liberalization and privatization of energy markets. Pressures in recent years have seen some—albeit limited—questioning of this role of the state and the development of a view that the state can intervene directly and facilitate the transformation of places that were previously told to make themselves adaptable to the requirements and priorities of mobile capital. This was beginning to contribute to a collection of national priorities where the tension is between trying to buttress the existing growth model and to explore new forms of low-carbon industrial interventionism. However, the new government's priorities signal a shift away from active intervention in partnership with urban and regional agencies to a simultaneously more localized approach through LEPs and stronger central control of more limited regeneration and green investment funds.

As we argued at the start of this paper, the MLP has primarily been concerned with national-level transitions and the role of the sub-national has been presented as one of primarily contributing to national transitions. Yet what the Greater Manchester example demonstrates is the complex territorial and national politics that coalesce within a city-region especially those of multiple, often contradictory, national priorities that shape and become incorporated in to place-based activities. National government engaging with cities is shorthand for a variety and multiplicity of relationships. Thus, who and what the city is in respect of transition is not reducible to place-based actors or institutions but needs to be understood through a multi-level, structural and processual politics of ongoing negotiation of priorities, possibilities and financial and knowledge resources. In times of the relatively stable reproduction of the city, these structures and processes may be fairly constant whilst when priorities shift radically—exemplified by a change of national government—they require effort and new organizations within which to organize action to such ends.

In Greater Manchester low-carbon activities were based on a narrow social "vision" of transition. This was primarily focused on a vision of economic costs, economic potential and the possibilities of urban low-carbon activities demonstrating how to achieve national targets. This "vision" was developed by exclusive coalitions with limited input from selected Greater Manchester social interests and through a national designation. In other words, the constitution of the vision was not part of a systemic participatory process of enrolling necessary and relevant others nor was it constituted as a space for struggle between competing potential visions and expectations. What followed from this was the development of three different intermediary organizational forms—CCA, LCEA, Energy Group—which although there were some overlaps in memberships were predicated on quite different geographies and priorities.

Each was also initially organized on short-term funding arrangements in a context of attempting to address longer-term transition in energy systems. This meant that vision and translation were not effectively interrelated and stabilized through organizational capacity and capability. They were also subject to struggle particularly in terms of the "economic" and "environmental" interests within the city-region—with the long established "economic" interests holding the balance of power in shaping a low-carbon agenda. This has had and will have significant implications for how national priorities are appropriated at the city-regional level and for the development of low-carbon energy capabilities. There is limited capacity to act but where there is capacity it is being mobilized largely in relation to an economic agenda rather than the wider incorporation of environmental and social priorities. It is capacity that is predicated largely, though not solely, on the symbolic representation of transition rather the serious enactment of a transition.

The early stages of making a low-carbon urban transition in Greater Manchester was predicated on a multiplicity of relationships and issues—that included public and private and national, regional, city-regional, local, supranational and business interests. The principal relationship was between national priorities and interests and a nascent city-regional governance framework that encapsulated different interests. Where capacity is being mediated and organized through the CCA this is being done through building relationships primarily horizontally and upwards with little downwards movement other than to select symbolically important projects and re-package them. This opens up the issue of for what purpose are governance relationships being built? Are they about re-constituting a status quo—that is the continuation of a form of urban economic growth—or about genuine transformation?

Environmental interests and those of the incumbent energy regime are involved but remain relatively peripheral. In that sense, the intermediary organizational forms mobilize a capacity to act that re-enforces the status quo of urban economic activity. There is not significant transformative capacity to re-configure and re-scale effectively the energy regime at the city-regional level—capacity to act is capacity to conservatively reproduce rather than radically transform. The dominant relationship is between the priorities of the strategic metropolitan level in Greater Manchester and national priorities and the ways in which they produce an "outward" looking Greater Manchester agenda, a symbolic agenda in the vein of an urban entrepreneurialism that has reconfigured its narrative to include low-carbon responses but where systemic transformation is much more difficult to find.

The transformative low-carbon agenda within Greater Manchester is much more difficult to locate other than implicitly through reference to projects and the issue this raises is whether what is being developed is largely a symbolic transition, primarily for the purposes of competing in the eco-economic race between cities rather than contributing to a transformative place-based transition. This gets to the very heart of "who" the city is. In the new governance arrangements it appears that the city, in this sense, is a strategic actor in the race between cities but with very limited operational capabilities. To put this another way, the strategic intentions of the city-region are largely disconnected from the city-region's civil society and a notion of its public sphere. What we are seeing is that in the balance of relationships a low-carbon transition in Greater Manchester is being driven by the national defining the city-regional and a narrow set of economic interests dominating the environmental.

6. Conclusions

In this paper, we have used research conducted in Greater Manchester to examine organizational contexts constituted for the purpose of transition to a low-carbon urban energy future. In this conclusion, we examine their wider implications and issues for further research.

First, we have sought to develop a better understanding of the dynamics of the relationships between a variety of national, regional, city-regional and local authority actors in addressing how low-carbon urban transitions are made, or more accurately how the language of transitions is mobilized by narrow coalitions of interest that do not result in genuine, radical transition but, rather, work to reproduce the economic status quo. In particular, we wanted to develop a more sophisticated appreciation of the ways in which urban low-carbon transitions are constituted—and the extent to which they are practically delivered—as metropolitan level responses to national priorities—such as carbon control, new industrial interventionism, a systemic long-term UK low-carbon transition, and innovation in relationships between national priorities and sub-national territories.

Second, our interest was in what shape these responses take, whether they are piecemeal or more strategic and systemic, the types of technologies implicated in responses, indeed whether these are primarily technological fixes or encompass cultural and behavioural change strategies or, indeed, both. But primarily what we were interested in was in the politics and organization of how such a response was produced and enacted, who was involved and who was excluded—or more specifically, what sorts of social interests and views were represented in constituting a response? In particular, we have sought to demonstrate that the relationship between city and energy system is not unproblematic. In a UK context, this has historically meant that the organization of energy regimes and those of urban coalitions of interests have often been organized at different scales. The implications of this are that conventional urban economic growth has been predicated on approaches that have not required a radical re-consideration of the energy regime. Energy regimes have usually supported rather than have been the specific focus of economic activity. A new set of ecological pressures have created the context whereby securing economic growth becomes much more tightly integrated with both the vulnerability and security of energy supplies and the wider capability to exploit the economic benefits of symbolic allusion to low-carbon urban transitions. In this sense, energy issues become more strategic in urban economic development priorities and strategies.

This shift creates an increasing set of pressures for reconfiguring socio-technical energy regimes at a city scale. In this respect, this is not just about technological artefacts at a city scale but also the social interests, institutions and actors, who seek to shape a city-regional energy regime. Where energy regimes are often organized nationally and regionally this means creating organizational contexts—that creates the necessary but not sufficient conditions—for the constitution of an urban energy regime that allows for social interests from the existing regime and city-regional interests and policy-makers to come together and communicate. In Greater Manchester, the additional complication was that city-regional governance structures were themselves being re-made with a central role for climate change and energy governance in doing so. In doing this we were keen to demonstrate the messy politics of these organizational responses, the dominance of national and economic priorities that were being advocated and the narrowness of the urban transition that might be produced if these intermediary forms can be stabilized.

Finally, the landscape pressures that shape urban transitions are the subject of considerable discussion among national and urban political and policy interests, researchers, community groups and activists. In particular, how the challenges posed by climate change and the geopolitics of resource flows can and should be addressed in an emerging age of economic "austerity" potentially has significant consequences for the shape of future urban transitions. In thinking about future research agendas and the mediation of social interests in low-carbon urban transitions, four issues would be worthy of further research. First, in what different ways are the current confluence of economic and ecological crises and pressures being used to influence low-carbon urban transitions? Are they being used to defend or challenge the status quo? Second, to what extent do these crises and pressures contribute to or constrain the possibilities for different notions of low-carbon urban transitions? Third, in different urban contexts to what extent is the mediating of low-carbon transitions a participatory process, how are these organized and with what consequences? Finally, must low-carbon urban transitions emanate from a national or city-regional level? What are the possibilities for grassroots and community initiatives to inform such transitions?

Notes

1. The chapters in a recent book (Bulkeley *et al.*, 2010) explicitly exploring the role of cities in low carbon transitions detailed some of the many different ways of thinking about the roles of cities in this respect—the city as a transition actor, as a contributor to national level transitions (Geels, 2010), through the lens of decision-making calculus (While, 2010), as constituted through multi-level governance coalitions of interest, the organisational cultures of urban transition (Hodson & Marvin, 2010a), but also through 'alternative' spaces within the city through which often marginalised voices seek to participate in low carbon transitions (Pickerill, 2010).
2. The LCTP claims that by 2020: more than 1.2 million people will be in green jobs; 7 million homes will have benefited from whole house makeovers, and more than 1.5 million households will be supported to produce their own clean energy; around 40% of electricity will be from low-carbon sources, from renewables, nuclear and clean coal; and the UK will be importing half the amount of gas that it otherwise would (DECC, 2009a).
3. We use the terms Greater Manchester and Manchester city-region interchangeably.

References

AGMA Executive Board (2009) *Statutory City Regions and Greater Manchester Strategy*, 27 February. Report to AGMA Executive Board.

AGMA & Manchester Enterprises (2008) *The Manchester Multi-Area Agreement: Our City Region's Proposal to Government.*

BERR (2007) *Energy White Paper: Meeting the Energy Challenge* (London: HMSO).

BIS/DECC (2009) *Low Carbon Innovation Strategy: A Vision* (London: HM Government).

Bourdieu, P. (1993) *Sociology in Question* (London: Sage).

Brenner, N. (2004) *New State Spaces: Urban Governance and the Rescaling of Statehood* (New York: Oxford University Press).

Bulkeley, H., & Betsill, M. (2003) *Cities and Climate Change: Urban Sustainability and Global Environmental Governance* (Routledge: London).

Bulkeley, H., Castan-Broto, V., Hodson, M. & Marvin, S. (Eds) (2010) *Cities and Low Carbon Transitions* (London: Routledge).

Climate Change Act 2008. Available at http://www.legislation.gov.uk/ukpga/2008/27/contents.

Coenen, L., Benneworth, P. & Truffer, B. (forthcoming) Towards a spatial perspective on sustainability transitions, *Research Policy.*

Committee on Climate Change (2008) *Building a Low-Carbon Economy—The UK's Contribution to Tackling Climate Change.* The First Report of the Committee on Climate Change, December, London.

Coutard, O. & Rutherford, J. (2010) Energy transition and city-region planning: Understanding the spatial politics of systemic change, *Technology Analysis and Strategic Management*, 22(6), pp. 711–727.

DCLG (2007) *The Review of Sub-national Economic Development and Regeneration: Sub-national Review* (London: HM Government).

DCLG (2008) *Explanatory Note on Taking Forward Proposals on City-Regions in PBR.* Available at http://www.push.gov.uk/pjc-090115-r06-sja-appendix_1.pdf.

DECC (2009a) *The UK Low Carbon Transition Plan* (London: HM Government).

DECC (2009b) *The UK Renewable Energy Strategy* (London: HM Government).

Delivering an Energy Group and Plan for Greater Manchester: An Update on Progress (2009) Report to Environment Commission, 10th September.

Deloitte (2008) *"Mini Stern" for Manchester: Assessing the Economic Impact of EU and UK Climate Change Legislation on Manchester City Region and the North West*, September.

Environment Commission (2010) *Work Programme Implementation Plan*, 20th January.

Geels, F. (2002) Technological transitions as evolutionary reconfiguration processes: A multi-level perspective and a case study, *Research Policy*, 31(8/9), pp. 1257–1274.

Geels, F. (2004) From sectoral systems of innovation to socio-technical systems. Insights about dynamics and change from sociology and institutional theory, *Research Policy*, 33(6/7), pp. 897–920.

Geels, F. (2005) Co-evolution of technology and society: The transition in water supply and personal hygiene in the Netherlands (1850–1930)—a case study in multi-level perspective, *Technology in Society*, 27(3), pp. 363–397.

Geels, F. (2010) The role of cities in technological transitions: Analytical clarifications and historical examples, in: H. Bulkeley, V. Castan-Broto, M. Hodson & S. Marvin (Eds) *Cities and Low Carbon Transitions* (London: Routledge).

Harvey, D. (2006) *Spaces of Global Capitalism. Towards a Theory of Uneven Geographical Development* (London: Verso).

HM Treasury (2008) *Pre-Budget Review*. Available at http://webarchive.nationalarchives.gov.uk/20100407010852/http://www.hm-treasury.gov.uk/prebud_pbr08_index.htm.

HM Treasury (2009) *Budget*. Available at http://webarchive.nationalarchives.gov.uk/20100407010852/http://www.hm-treasury.gov.uk/bud_bud09_index.htm.

Hodson, M. & Marvin, S. (2010a) *World Cities and Climate Change* (Maidenhead: McGraw-Hill).

Hodson, M. & Marvin, S. (2010b) Can cities shape socio-technical transitions and how would we know if they were? *Research Policy*, 39(4), pp. 477–485.

Hodson, M., Marvin, S., Bulkeley, H. & Castan-Broto, V. (2010) Conclusion, in: H. Bulkeley, V. Castan-Broto, M. Hodson & S. Marvin (Eds) *Cities and Low Carbon Transitions* (London: Routledge).

Infrastructure Cross Commission Working Report to Environment Commission (2009), 10th September.

Manchester Independent Economic Review (2009) *Manchester Innovation Investment Fund*.

Manchester Statutory City Region Agreement (2009), December.

Massey, D. (2005) *For Space* (London: Sage).

Minutes Environment Commission (2009), 10th September.

Pickerill, J. (2010) Building liveable cities: urban Low Impact Developments as low carbon solutions? In: H. Bulkeley, V. Castan-Broto, M. Hodson & S. Marvin (Eds), *Cities and Low Carbon Transitions* (London: Routledge).

Prosperity for All: The Greater Manchester Strategy (2009), August.

Report to Environment Commission on Climate Change Agency (2009), 14 May.

Report to Environment Commission on Energy Planning in Greater Manchester (2009).

Report to Environment Commission on Environment Commission Away Day—8 June 2009 (2009), 14th May.

Report to Environment Commission on Environment Commission Work Programme: Commissioner Roles (2009), 2nd November.

Report to Environment Commission on Low Carbon Economic Area (2010a), 20 January.

Report to Environment Commission on Low Carbon Economic Area (2010b), 18th March.

Report to Environment Commission on Low Carbon Economic Area and CCA Update (2009), 2nd November.

Report to Environment Commission on the Transition to a Low Carbon Economy (2009), 16th July.

Report to Planning and Housing Commission on Next Steps for the AGMA Decentralised/Low Carbon Energy Study (2010), 17 March.

Shove, E. & Walker, G. (2007) CAUTION! Transitions ahead: Politics, practice, and sustainable transition management, *Environment and Planning A*, 39(4), pp. 763–770.

Smith, A., Voβ, J. P. & Grin, J. (2010) Innovation studies and sustainability transitions: The allure of the multi-level perspective, and its challenges, *Research Policy*, 39(4), pp. 435–448.

Späth, P. & Rohracher, H. (2010) Energy regions: The transformative power of regional discourses on socio-technical futures, *Research Policy*, 39(4), pp. 449–458.

Voß, J.-P., Bauknecht, D. & Kemp, R. (Eds) (2006) *Reflexive Governance for Sustainable Development* (Cheltenham: Edward Elgar).

While, A. (2008) Climate change and planning: Carbon control and spatial regulation, *Town Planning Review*, 79(1), pp. vii–xiii.

While, A. (2010) The carbon calculus and transitions in urban politics and urban political theory, in: H. Bulkeley, V. Castan-Broto, M. Hodson & S. Marvin (Eds) *Cities and Low Carbon Transitions* (London: Routledge).

Local Demonstrations for Global Transitions—Dynamics across Governance Levels Fostering Socio-Technical Regime Change Towards Sustainability

PHILIPP SPÄTH* & HARALD ROHRACHER**

*Institute of Forest and Environmental Policy, University of Freiburg, Freiburg, Germany, **University of Klagenfurt, Inter-University Research Centre on Technology, Work and Culture, Graz, Austria

Abstract *Which role do spatial dimensions play in the transformation of socio-technical regimes, in particular the energy system, towards more sustainable configurations? Concepts such as the multi-level perspective on socio-technical change have not given sufficient attention to space and place so far. We develop our considerations around the case of an "Energy Region" in Austria where people try to bring about a substantive shift in their "local" energy supply structure and have the ambition to contribute to a "general" transition towards sustainable energy systems. However, if this ambition is to stand the test of reality, what are the mechanisms and processes through which regional governance can have a broader impact on the transition of the energy system? What are the resources it can draw upon? What are the linkages with other governance levels? We investigate in detail how one regional showcase for the feasibility of a non-fossil, sustainable energy system was set up in Murau, a remote, alpine district of Austria. Starting from the multi-level framework for the modelling of niche-regime interaction, we put particular emphasis on the formation of discourse coalitions and dynamics of multi-level governance. Our findings support the view to pay considerably more attention to the interplay of local and non-local discourses and the dynamic relations between local initiatives and non-local networks which can provide specific opportunities for the legitimization and entrenchment of alternative socio-technical configurations.*

1. Introduction

In their attempts to explain (and support) regime change, scholars of the system innovations approach to socio-technical transitions highlight the importance of "experimentation and learning" in niches. The approach of "strategic niche management" in particular stresses the need "to create alternative configurations" in niches. It assumes, that such "variation" in protected spaces has the potential to exert sufficient pressure to fundamentally change or ultimately replace entrenched systems (Hoogma *et al.*, 2005).

Empirical and historical studies have shown, however, that the existence of technological alternatives, even those clearly superior to dominant configurations, is not a sufficient precondition for regime change (David, 1985). Transition scholars point out that pressures at a macro-social level, shocks (e.g. oil supply shortages) and a destabilization of the socio-technical regime (e.g. political pressure to reduce greenhouse gas emissions) are equally needed to induce transformational change (Schot & Geels, 2008). Nevertheless, even in this more differentiated view, niche-regime interactions are a core element to understand the dynamics of transitions to sustainable socio-technical regimes. What is missing in this respect, is a more complete understanding of how socio-technical niches are aggregated and combined to have an impact at the regime level. "In addition to identifying opportunities for niche-regime connections", Smith (2007a, p. 437) writes, "we need to understand the connecting processes how these reconfigure developments in niche and regime". For any alternative to become a challenge for the entrenched regime, the mobilization of heterogeneous actor networks who "build up credibility" around alternative configurations is a crucial prerequisite of its institutional embedding.

We believe that regional or municipal policies provide such a linkage between niche and regime levels. These policies often do not aim primarily at particular niches and technological experimentation. Rather they demonstrate the feasibility of new configurations of niche technologies embedded in modified social and political practices at a regional level. A typical setting for the demonstration of such alternative configurations are "model regions" (Späth & Rohracher, 2010). Ideally, these regions should be representative for at least a larger group of regions to allow for up-scaling of experiences, though they are often selected on the basis of their providing conditions which increase the likelihood of success of the model projects.

Advantageous conditions for such variants of regime constellations can be linked to certain "places". A particular region or city can, for example, be collectively perceived as predestined for an energy transition (Späth & Rohracher, 2010). Other crucial qualities correspond to clearly delineated territories,[1] for example, administrative institutions and line-bound infrastructures. The variation of these conditions in space results in a fine-grained patchwork of possibilities in which dynamics evolve that can contribute to large socio-technical transitions.

Researchers interested in socio-technical transitions towards sustainability have only recently started to conceptualize such geographies of transitions (Coenen *et al.*, 2010; STRN Steering Group, 2010). These efforts can also build on an increasing interest, in both political practice and social research, to understand the role of cities and regions as loci for the development "and" exemplary implementation of alternative regime configurations. The battle over the energy future has predominantly been fought at the national level where the formal competencies for energy policy mostly lie. Since the 1980s, however, municipalities and regional bodies have increasingly contributed to climate and energy political debates (Bulkeley, 2005; Bulkeley & Betsill, 2005; Hodson & Marvin 2009; Monstadt, 2007). While this can be understood as a process of regionalization, local actors often frame their activities as part of a global fight for particular socio-technical transitions. In line with this motivation, cities and regions set up various networks such as the Cities for Climate Protection Campaign, the Climate Alliance of European Cities, the Aarhus Convention and many more, and strategically join forces and exchange experiences in a supra-national space (Bulkeley, 2005).

Contemporary transition literature characteristically refers to the Multi-Level Perspective (MLP), which distinguishes socio-technical niches, regimes and landscape as three "levels of structuration". This framework is particularly helpful in explaining the aspects of "variation and selection". Yet the aspect of "building up credibility" for given alternative configurations in "model regions" calls—as we will show—for a more explicit analysis of the dynamics that can result from strategic behaviour in systems of "multi-level governance". We need to complement our analytical framework with a heuristic that allows us not only to understand how experimentation for the sake of "variation" but particularly also how "implementation for the sake of demonstration" is influenced by and dependent on spatially differentiated opportunity structures. For both kinds of activities their specific embedding into particular places is highly relevant, and cities or regions seem to be crucial entities in this respect.

Feasibility demonstrations in model regions or cities can lend credibility to a discourse around particular alternative configurations. This may help to legitimize institutionalization processes beyond the model regions, for example, by influencing national research funding priorities or other institutions that previously supported the entrenched regime. "Model regions" therefore often are crucial in the strategies of national or international networks organized around alternative socio-technical configurations (Späth & Rohracher, 2010).

We empirically observed that social actors differentiate and localize elements of both "local nodes" and "global networks"—to use the terminology of Smith (2007a). This is done on various spatial scales, from single sites of energy generation or consumption to a worldwide scope. Strategies of cross-level fertilization between spatial scales and/or governance levels are of particular importance for the process of building up credibility and authority for an alternative configuration.

Our analytical perspective therefore needs not only to combine the evolutionary "multi-level framework" of niches, regimes and landscapes with a discourse analytical approach to policy making, which includes the aspects of discourse structuration and discourse institutionalization as conceptualized by Hajer (1995). We should furthermore take into consideration the strategic games that are played across governance levels and should therefore draw on the literature on Multi-Level Governance (Smith, 2007b).

Many examples of strategically designed processes of feasibility demonstrations in "Model Regions" have been observed over the last few years in the energy domain. Particularly in Germany and Austria, programmes for the competitive funding of model energy regions have been implemented under the auspices of federal ministries. Many of the model regions funded under these schemes aim at some sort of "energy autarky" or 100% local net-supply from renewable energies (Radzi, 2009).

In Germany, a considerable number of regions are on their way to become "bio-energy region" or "100% region", and hundreds of villages are aiming to become "bio-energy villages" (Moser *et al.*, 2009). This recent development is still awaiting an in-depth analysis in social scientific terms. Previous programmes for the competitive funding of region-specific development initiatives, such as "RegionenAktiv", have been extensively analysed with regard to typical dynamics of network building and socio-economic and institutional success factors (Böcher *et al.*, 2008; Elbe *et al.*, 2007). Arthur Benz conceptualized consequences of this type of "governance by competition" particularly in multi-level governance systems (Benz, 2004).

In Austria, a public climate and energy fund spends about €150 million per year on climate protection projects and energy research and promotes its funding of energy model regions as a "flagship activity". Besides particular measures to increase energy efficiency and the use of renewable energy in those regions, the training of network builders has now become a major field of activities that the fund supports.

In this paper, we describe how one such regional showcase for the feasibility of a non-fossil, sustainable energy system was set up. We use the case of the remote, alpine district of Murau (roughly 30,000 inhabitants) in Upper Styria, Austria, with the city of Murau (roughly 2100 inhabitants) as its only town to explore the role of strategically exploited dynamics between various levels of governance. An initiative to achieve "energy autonomy" in this district was supported by a broad range of actors.[2] Despite Murau's specific geographical and socio-economic situation it can serve as a generalizable example for the characteristic mechanisms and dynamics at work in such model regions. The case is thus representative of a broader group of socio-spatial framings of niche-regime interactions in sustainability transitions.

In an earlier paper we analysed the case of Murau with a particular focus on the role of guiding visions in coordinating action (Späth & Rohracher, 2010). In this paper, we focus on actors' strategies within an interplay of local and non-local discourses and on the extent to which dynamic interdependence between local initiatives and non-local networks can provide opportunities for the legitimization, institutionalization and therewith entrenchment of the alternative configuration in various domains and across governance levels.

The following section starts off with a summary of relevant conceptualizations from the literature on socio-technical change and systems innovation, which we then complement with elements from the literature on Multi-Level Governance. In the third section, we illustrate the relevance of our conceptual account with observations from our case study region in Upper Styria, Austria. Against this empirical background, we discuss the role of cities and small regions in socio-technical transitions (Section 4), before we finish by summarizing our conclusions.

2. Combining Theoretical Approaches on Socio-Technical Transition, Discourse Institutionalization, Space and Multi-level Governance

2.1 *Socio-Technical Regimes, the Management of Niches and the Role of Cities and Regions in Building Up Authority for Alternative Socio-Technical Configurations*

The acknowledgement of severe "sustainability problems" such as climate change and depletion of energy resources has fuelled a renewed interest in the way in which "socio-technical systems" change over time and to what extent and in which ways such processes of change can be governed. Energy systems, for example, are characterized by many stabilizing interdependencies and path dependencies. However, as history shows, this does not exclude phases of major systemic change. Technology studies deal with the inseparability of social and technical elements in socio-technical configurations such as the energy system and provide insights in the obduracy of such configurations. On this basis dynamics of socio-technical change processes and opportunities and constraints for governance of such processes are explored. The conceptual literature on "socio-technical systems" and the MLP (see, e.g. Geels, 2004; Rip & Kemp, 1998) have been found to provide useful heuristics for structuring the usually messy accounts of complex system dynamics. The regime perspective is helpful for thinking about energy system transformations in terms of stability of existing configurations, as a function of the various social and technical elements that have to come together to cause a regime shift and, not least, in terms of the integrated and long-term policy strategies needed to shape such regime shifts (Rohracher & Späth, 2009).

The more prescriptive "governance" approach of "transition management" even suggests that the identification of concurrent developments at various levels can help actors to identify such phases of increased probability of regime change and its possible directions and thus provides particular opportunities for strategic support of such developments (Grin *et al.*, 2010; Loorbach, 2007, 2010). At the core of the so-called MLP on socio-technical change is an evolutionary concept of variation at the micro-level (i.e. the "breeding" of new technologies within confined technological niches or market segments) and of selection at a meso-level (dominant socio-technical "regimes"), embedded in the broader context of the "socio-technical landscape". This landscape encompasses cultural norms, values and broader social structures which only change very slowly. In contrast to its biological counterpart of variation-selection, transition literature considers under certain conditions an "overthrowing" of a selection environment by variation processes or a so-called regime shift. Moreover, the multi-level model is founded on an understanding of different (and separable) temporal dimensions of change—from short-term processes at a micro-level to the "longue duree" of changes at the macro-level of socio-technical landscapes.

At the meso-level of the "socio-technical regime", socio-technical configurations are temporarily stabilized and supported by a rule set or "grammar" that structures the socio-technical co-evolution process. The regime incorporates the mutually reinforcing technological and institutional structures of specific domains, such as the energy system, and is characterized by a resistance to change (which, for example, may cause promising new technologies to fail). Some aspects of this conceptual framework still require further clarification: How niches and regimes exactly influence each other and due to what generic "mechanisms" niche developments can aggregate and link into regime change are still not very well understood (see, e.g. Geels & Schot, 2007; Smith, 2007a; Smith *et al.*, 2005).

The activities within energy regimes, for example, are coordinated through various rules—not only at a regulative/legislative level, but also by cognitive (and normative rule sets (paradigms and cognitive frames, values and expectations) (Geels, 2004). What has not yet been sufficiently addressed is the extent to which these regulatory, cognitive and normative rule sets can, or usually do, vary in relation to the particular conditions given in different localities.

The approach of Transition Management as well as the older and less general approach of strategic niche management (SNM) seeks to develop strategies for the establishment of "niches", i.e. "protected spaces" for new technologies to develop which can potentially challenge entrenched regimes. In line with their quasi-evolutionary understanding of socio-technical innovation, these approaches emphasize the importance of "local" experiments in "protected spaces" for the development of "variation" in socio-technical configurations. The availability of a high number of alternative configurations increases the possibility that one of them will finally develop a sufficient fit with the changing selection environment and challenge and destabilize the entrenched regime. From that perspective, the probability of regime transition is mainly a function of the variety of socio-technical configurations that are experimented with in various niches. While the availability of alternative configurations may be a necessary precondition for regime transformation, other factors such as the "building up of authority" for already well known alternative configurations could be just as crucial.

As mentioned earlier, "niches" are conceptualized as "spaces" that are protected from certain economic selection pressures. However, the role that physical and social space and socially constructed places such as cities or regions can play in such processes, has so far been largely neglected in the transition literature (Hodson & Marvin, 2009). Given the prominent public role of cities in fighting climate change and the importance of regions for innovation and technological change (Truffer, 2008) this blind spot neglects an important dimension of socio-technical change and related potentials for governing such transformation processes. Only a limited number of studies have been dealing with climate policy and the governance of energy infrastructures in cities from a socio-technical point of view (Bulkeley & Kern, 2006; Coutard & Guy, 2007; Graham & Marvin, 2001; Guy & Marvin, 1996; Hodson & Marvin, 2009; Monstadt, 2007) or in regions (Smith, 2007b; Späth & Rohracher, 2010).

Coenen *et al.* (2010) discussed possible conceptualizations of experiments in "particular environments", i.e. "localities". building on earlier attempts to extend the SNM approach by differentiating local and global aspects of "niches" (Geels & Raven, 2006). Along similar lines, the network for sustainability transitions research, in its first research agenda explicitly calls for more attention on "the spatiality of transitions" and for research on the role of cities and regions in transition processes (STRN, 2010).

In our understanding, cities and regions have to be located outside or rather across the levels of niches, regimes and landscape. Cities and regions may, for example, be steeped in the energy regime but at the same time can serve as spaces where important partial transformations can take place. While the notion of socio-technical regime conveys a sense of homogeneity, regimes are in fact far from this and considerable spatial variations exist. The transition literature predominantly focuses on external micro- or macro-level forces that destabilize and transform well-established and largely homogenous socio-technical regimes (i.e. transformative momentum is built up by niches). In contrast, we want to highlight the existence and importance of regime inconsistencies and variations. The regime level is usually conceptualized as comprising the various features of a complex socio-technical system, ranging from the supply side to end-users. The current energy regime, for example, is essentially characterized by a plethora of institutions, actors and interests aligned to large-scale and fossil-fuel-based systems that are partly of global reach. The regime should in our view therefore be conceptualized as of global or at least national scope.[3] An energy "transition"—understood as a radical, systemic and transformational change—would therefore have to take place at this trans-local and in consequence trans-national scale. There can, however, be significant variations in regimes between nations, but also between regions or cities. There is also room for manoeuvring at these different levels to strategically transform parts of the regime. Small deviations at a local level, how marginal they might appear at first, could play a cumulative role with other factors and collectively contribute to destabilizing the regime. As we will show with the following case study, local actor networks do intentionally create and support such "deviations" from the principle characteristics of a regime. Such variation is certainly constrained by many factors, but can have an impact on national or even international discourses and policies of regime change.

In this article we are particularly interested in the potential of such "regime deviations" to "lend legitimacy" to visions of more sustainable socio-technical regimes and related processes of learning, actor-alignment and institution building. We deliberately focus on socio-technical configurations at an intermediate geographical scale, i.e. in cities and smaller regions. This is done with the expectation that we uncover particular conditions and dynamics which might be just as necessary for socio-technical transitions to gain momentum as those at the national and international level. Cities and regions are often large enough to incorporate at least some of the systemic properties of existing regime structures (typical supply structures, relations of supply and demand, etc.). At the same time, they can be small enough to exploit the advantages of proximity for the creation of new actor networks, discourses and institutions for alternate socio-technical configurations. Thirdly, they often have well-established governance structures to coordinate change processes (not only city or regional governments, but often also well-organized civil society organizations and networks involved in urban or regional governance).

Regime variations at lower spatial scales are thus important resources for multi-scalar processes of regime transitions towards sustainability. This multi-scalar dynamic of regime change is not the same as the multi-level innovation dynamics described in the MLP. However, it may help us understand how aggregates of niches can be provisionally stabilized in particular places and how that can build up potentially influential transition momentum at the sub-regime level.

We have argued elsewhere that it is necessary and possible to complement the analytical perspective of socio-technical transitions with a heuristic that can help us understand the discursive dynamics increasing or diminishing the impact of particular socio-technical visions (Späth & Rohracher, 2010). In this regard we follow the discourse coalitions approach developed by Maarten Hajer (Hajer, 1995). This approach analyses the striving for "discursive hegemony" by building coalitions, mobilizing various resources and particularly by accumulating "authority" for preferred storylines and policies.

> The dynamics of this argumentative game is determined by three factors: credibility, acceptability, and trust. Credibility is required to make actors believe in the subject-positioning that a given discourse implies for them and to live by the structure positionings it implies; acceptability requires that position to appear attractive or necessary; trust refers to the fact that doubt might be suppressed and inherent uncertainties might be taken for granted if actors manage to secure confidence either in the author (whether that is an institute or a person), e.g. by referring to its impeccable record, or in the practice through which a given definition of reality was achieved, e.g. by showing what sort of deliberations were the basis of a given claim. (Hajer, 1995, p. 59)

This perspective sheds light also on discursive dynamics that build on the social construction of "geographical scales" or on institutionalized "levels of governance". We will see how this can help us with understanding the interplay of local nodes and global networks in recent attempts to bring about an energy transition.

Finally, dynamics of multi-level governance are gaining ever more importance in contemporary political sciences. Building on classics such as Benz and Eberlein (1999) and Bache and Flinders (2004), Marks and Hooghe (2004) offered a typology to better understand empirical trends in the ways that levels of governance are constructed and how they are related. The highly formalized, nested hierarchies of governance levels that seem to have dominated the scene in federal systems thus far, (type I multi-level governance arrangements) are increasingly complemented by much more fluid and complex network-type governance arrangements, which loosely integrate a more heterogeneous set of actors (type II multi-level governance arrangements). At the same time Morgan, responding to the "new regionalism" debate, described how "regions" are constituted and how the "regional scale" relate to other scales in what he calls the "multi-level polity" of the European Union (EU) (Morgan, 2004).

Problems of environmental governance are recognized to notoriously impact on multiple geographical scales. The governance of these problems therefore particularly calls for Multi-Level Governance approaches both in practice and in its analysis (Knill & Liefferink, 2007; Mirumachi & Van Wyk, 2010). Bulkeley (2005) stimulated an intense debate about the role of cities' coordinated initiatives to combat climate change and how such initiatives transcend conventional understandings of multi-level governance (Bauriedl, 2009; Braun & Santarius, 2008; Bulkeley & Betsill, 2005). Drawing also on multi-level governance literature, Smith (2007b) discussed the issue of policy learning between regions and central government in the case of renewable energy support in the UK. Wissen (2009) pointed out how the "scale debate" in radical geography, with its process-based understanding of transformations in statehood, helps to "overcome the limitations of more static concepts like multi-level governance". He demonstrated using the case of property rights on genetically modified organisms, that the production of new spatial scales of the state can be an important element in the reorganization of control, e.g. over nature.

Combining such a perspective on the activity of scaling with an analysis of the formation of discourse coalitions and respective networks shall now help us to assess the potential of local activities to influence transformative niche-regime dynamics.

We are only starting to develop a shared analytical framework to help us better understand the politics of scale involved in socio-technical transitions. Our analysis above points to some critical issues which should be followed up in further research. Firstly from an actor oriented perspective, we need to carefully study in what arenas the various actors interested in challenging incumbent regimes are getting active and what discursive strategy they employ. It is probably revealing what kind of social network they (aim to) establish and which positions they take up themselves. The spatial connotations (local, regional, provincial etc.) connected with each actor, each arena and each statement are very important to take into consideration. Particularly in the struggle about the establishment of new political arenas, notions of scale might be important. Secondly, from a discourse-oriented perspective we need to study the building up of authority for certain storylines, the struggle for dominance and the institutionalization of discourses by discourse coalitions. Thirdly, from a more system-oriented perspective, we need to take into consideration the trajectories of established socio-technical systems and the forces that can potentially overcome the tendency for growing regime stability. Among these, the usually slow changes at the landscape level and the potential of niche experiments to accumulate are certainly important. In addition to these, we argue, we also need to put greater emphasis on the vulnerability of regimes resulting from challenging conditions in particular places and territories.

3. The District of Murau: Driven by "The Energy Vision"—Driving the Grand Energy Transition?

Over the last few years, actors in various cities and regions have developed ambitious objectives with regard to fundamental changes in local energy infrastructures and even consumption patterns. On one hand, these efforts are motivated by hopes to improve the environmental and economic impacts of the energy flows "locally". Responding to local conditions such as specific actor configurations and economic conditions, specific strategies are developed to change the local energy system. At the same time, however, these local initiatives are often linked to more "generic, non-local visions" of a transition towards sustainable energy systems at a national or global level.

Murau is a rural district in Upper Styria, Austria. It is located in a peripheral alpine valley, covered largely by forests. It is relatively sparsely populated, with approximately 31,000 inhabitants. The decrease in the local population over the last decade has exceeded the Styrian average (−2.4% in total from 1991 to 2001). The greatest potential renewable energy source comes from the considerable stock of woody biomass from the mostly privately owned forests.

3.1 *The Formation of a Local Agenda and Network*

In 2003, the head of the regional Energy Agency of Upper Styria, together with a professional facilitator of participatory processes, launched an initiative to develop and implement the "Energy Vision Murau". The process was initiated by bringing together a small circle of energy activists and drew from the concept of "Energy Regions", which guided similar regional initiatives in Austria since the early 1990s (Alber, 2009; Späth *et al.*, 2007). They developed the idea of establishing a process for broader participation in order to kick-off, facilitate and coordinate the implementation of various projects for renewable energy and energy efficiency in the region. A central aspect of their idea was to embed the renewed interest in biomass-heating into a more comprehensive approach of energy system change and regional development and to create synergies between, and legitimacy for, a variety of agendas and projects.

3.1.1 *A "guiding vision" for the district—the "Murau Energy Objectives for 2015"*[4]. These initiators invited organizations, companies and citizens from the region to a series of workshops. In the beginning 30 people, mostly representing local energy related organizations participated in the process. The workshop included gathering participants' ideas about how their region-specific context related to broader visions and objectives for sustainable energy futures, regional development and climate protection. The discussions revolved around storylines like "by using fossil energy we transfer our money to the sheiks" and "there is enough biomass in our woods for our full heat and electricity supply". All participants had a chance to contribute to the debate and moderators made sure that priorities were selected in consensus and reflected the willingness of participants to actively contribute to the implementation of objectives. These priorities were then cast into a sketchy vision of the regional socio-technical future: that of "energy autonomy". Specific objectives were also identified for the district, to be realized by 2015[5]:

- The district of Murau is energy autonomous with regard to heat and electricity, i.e. a positive balance for renewables in primary energy flows has been achieved.
- Prospering regional economic circuits of energy production and consumption have been established and a surplus of economic value is created by a net export of energy carriers.
- A high level of public awareness concerning the need for an energetic circular flow economy ("energetische Kreislaufwirtschaft") has been achieved (especially among pupils).

Measures to achieve these objectives were identified and prioritized. For the most important measures, permanent working groups were established who developed specific strategies focusing on, for example, wood-fired district heating systems, solar heating systems for private homes, projects for renewable electricity production and ways to improve the energy efficiency of buildings.

3.2 *Drawing on Local Resources and Specificities*

This agenda relates to two different discourses: A green, environmentalist discourse that promotes, for example, the de-carbonization of the energy system and a more materialist discourse on economic development, which primarily promotes local economic development and the regionalization of income and employment. Merging these two generic discourses and contextualizing them into one local energy political agenda allowed the local actor network to extend to much larger parts of the population than had been possible with a more narrow (e.g. purely environmentalist) agenda. Emphasis was thus put on the assessment of regional potentials for value creation by bringing regional, renewable energy sources into use, with biomass traditionally playing an important but not exclusive role in many alpine or agricultural regions of Austria.

The district with its small and dispersed population (the district town of Murau is home to only 2100 inhabitants) and limited industry does not allow for the profitable operation of a gas supply network and therefore is partly "off the radar" of the major incumbent actors in fossil-fuel provision. These specific conditions (location off the gas grid, huge biomass potential) and opportunities for mobilization (convergence of environmentalist and economic agenda, forest owning farmers, tradition of biomass use) have been intentionally exploited. Local natural and economic specificities hence coincided with a discursive opportunity and its strategic use in forming something like a "discursive niche". This can explain why the initiative and its new "paradigm" faced such weak resistance and was soon able to dominate the local discourse.

In terms of "social relations", the district of Murau is home to a relatively close-knit community, with frequent interactions between people, due to a high importance placed on community activities, associations, family ties, etc. On the other hand, the district accommodates enough enterprises, associations, ambitious politicians, etc., to make up for a sufficiently "heterogeneous" actor network in command of all resources required for a significant impact on local investments (development funds, media relations, etc.)

The network also gained cohesion from the catchy "ideographs" (van Lente, 2000) of "energy autarky" and "self-reliance", which, for historical reasons, has always seemed to be particularly attractive to Austrian farmers. For the activists in the region, this objective is not at all related to romanticized ideas about regional detachment, nevertheless, it was welcomed that such connotations exist in the broader discourse.

Referring to the role of programmatic ideas in policy making, Cambell stated that

> the probability that a programmatic idea will affect policy making varies in part according to the extent to which it provides clear and simple solutions to instrumental problems, fits existing paradigms, conforms to prevailing public sentiment, and is framed in socially appropriate ways. (Campbell, 1998, p. 399)

If the framing of problems and public sentiments varies from one local context to another, so too will the consideration of social appropriateness . One important differentiation in this socio-cultural space might be that a programmatic idea might develop a certain momentum more easily in small spaces with relatively strong cultural homogeneity.

"Institutional voids", emerging in certain places or territories might also provide supportive conditions for the development of new discursive arenas in which alternative energy political agendas can be negotiated. We hypothesize that such a void occurs more often at local levels than at more tightly contested provincial or national levels:

> Characteristically, these new spaces of politics initially exist in an institutional void: there are no pre-given rules that determine who is responsible, who has authority over whom, what sort of accountability is to be expected. Yet as politics takes place between organizations, all people bring their own institutional expectations and routines with them. And, as different participants follow their own "logic of appropriateness" (...), politics in new political spaces is never only about content, but inevitably also about the rules of the game and a dynamics of credibility. (Hajer & Wagenaar, 2003, p. 9)

The process of negotiating objectives and jointly organizing their implementation in Murau has actually occurred "in between" the realms of administration, major associations, enterprises and other stakeholders in an inter-organizational network. This resulted in actors, who would normally be in disagreement, reaching consensus in a joint agenda. Another characteristic in local configurations supportive of discursive hegemony is that a local population can often be very familiar with and accepting of certain energy experts due to their long-standing work in the region.

3.2.1 *The formation of a heterogeneous network in order to access various power resources.* The aim of the initiators was to mobilize various actors in order to gain access to particular resources under their control: the owner of a local company installing heating systems, a former mayor, the Secretary General of an agricultural association at regional level. Besides their professional knowledge and competence to articulate expectations with regard to a future energy system, they also added legitimacy by their reputation and were able to influence key decisions because of the positions they held in several societal domains. Politicians and officers of the regional and municipal administrations, for example, translated the joint vision into decision-making processes on public investments as well as into publicly funded awareness campaigns. Local business people jointly and individually adopted strategies explicitly in support of the objectives and promoted them to their customers. The involvement of well-respected business people was actively sought after since they lent credibility to the economic objectives and "rationality" of the process. The initiators of the Murau Energy Vision and the members of this "innermost circle" consequently describe themselves as mediators between the spheres of government, enterprises and "the public".

3.2.2 *Redundant institutionalization at many levels of regional governance.* The Energy Vision objectives have meanwhile been given the formal status of binding priorities for economic development at a regional level. These institutionalizations benefited from the fact that the head of the regional energy agency is not only a key actor in the Energy Vision process but also an officially appointed manager of the regional EU-LEADER-programme and at the same time involved in development planning on the NUTS-3 level—the broader region of Upper Styria West, inhabited by roughly 105,000 people. As a result, the objectives and priorities of the Energy Vision have been recognized as a guideline at "five" different regional levels: single municipalities, small planning associations of 4–10 municipalities (regionext), LEADER-Regions (equivalent with district level), NUTS-3 region and province ("Bundesland"). This success in institutionalizing the energy transition agenda in rural development at so many levels of governance was a heavily discussed strategy as it produced a high work load for the facilitators of the process, and was criticized to be redundant and stretching the available resources too much.. In retrospect, however, the managers were very glad to have chosen this costly path, as the dynamics at the various levels reinforced each other and contributed to making the setting of ambitious energy objectives a social norm across the larger region (Späth, 2008). The distinctively many regional governance levels that have developed in Styria and the various programmes for regional development funding, most importantly based on EU structural funds, together constituted an important part of the locally specific "opportunity structure" for regional networking and capacity building.

The process leading towards the "Energy Vision" of Murau has in particular been identified as a champion and leading example among regional development experts in Austria and even gained recognition abroad. The central actors in this process also contributed to setting up similar processes in other parts of the province (e.g. Eastern and Western Styria) and presented it to nation-wide audiences (e.g. in the contexts of an R&D programme and of the Austrian Energy and Climate Fund) and to international audiences (e.g. in the context of the EU LEADER+programme).

3.2.3 *The intermediation between "local node" and "global network".* Those actors responsible for strategic coordination played an important role in the strengthening of the Murau Energy Vision and initiative. They not only motivated selected stakeholders to participate, but also communicated the initiative via local, provincial and national media, through personal communication and as a "best practice example" in international (particularly European) circles. The director of the regional energy agency in particular had an important position within this network: by interest, training and profession, he is an expert in alternative energy projects and the broad discourse on "energy change" (Energiewende). He was involved both in debates concerning the economic development in the region, as well as discussions at national and EU-level on the design and management principles of funds for regional development. He was therefore able to bring in ideas and knowledge from broader discourses at national and international levels and to link the local network with respective networks of national and international scope (e.g. the regional development community, the international network of energy agencies, various EU-research projects etc.). Consequently, he was instrumental in mobilizing crucially important knowledge, funds and important contacts at multiple governance levels. From the perspective of the local network, he had not only a "gatekeeper position", but enabled the network development by crucial "boundary work" (Wenger, 2000).

3.2.4 *Beyond technical experimentation: creating a showcase for a new socio-technical configuration.* The consideration of ways to realize the objectives also led to local R&D efforts including a designated truck for delivering wood chips, the development of new logistical infrastructures like a regional biomass exchange and of new services related to biomass-heating in private homes. This improvement in technical solutions and business strategies went beyond a mere adaptation of ready-made solutions to regional circumstances. Some R&D projects were initiated, funded by national and EU R&D programmes, and this again contributed to the building up of a base of knowledge, skills and ideas that were instrumental in the energy vision initiative. The mobilization of funds for R&D in connection with the energy objectives also added to the reputation of the initiative as being scientifically sound, innovative and economically promising. This activity was also perceived as crucial in the setting up and promotion of an exemplar project that demonstrated, not so much technological feasibility, but the socio-economic feasibility of the idea of a deprived district freeing itself from the chains of energy dependency—by a combination of proven technology, changed business models and consumption patterns. This local process of becoming an "energy region" was meant to challenge the dominant energy regime mainly by translating ambitious visions of a sustainable energy future into rather mundane regional practices. It was explicitly set up to demonstrate that such transformations need not lead to serious social disruptions but may even benefit the regional economy. Whether this is realistic or not, the "show case" of a district aiming at energy autarky suggests a perspective for a potential "up-scaling" to a national or international level.

3.3 *Local Demonstration Activities Gaining Support from National and International Developments*

Such local initiatives can be a useful working example for proponents of change involved in discourses, e.g. at a national level who are otherwise caught up in interest-driven disputes about targets and their implementation. Therefore such proponents actively support the local experiments by giving them opportunities for (self-)promotion and some minimal funding. Right from the first proclamation of the energy objectives, the initiative received positive coverage in the local and regional press. In 2004, they participated in a competition called "Energy Regions of Tomorrow", which was proclaimed by the Federal Ministry for Transport, Innovation and Technology, and was one of nine initiatives to win an award. By way of this competition and by funding the further development of the "Murau Energy Vision" as an accompanying measure within its R&D-programme "Energy Systems of Tomorrow", this Ministry gave the Murau Energy Vision and other initiatives some official credibility.

What the initiators experienced as another breakthrough was the winning of the "Energy Globe Award" at both the Styrian and national levels in 2004. This glamorous price, the award of which was broadcast on national television, drew much attention to the region. It was also welcomed by the regional population as it entailed the Murau Energy Vision representing Austria in the competition at the European level.

Such successes attracted further regional and specialized press coverage and also coincided with a general increase in interest in energy and climate issues. Some activists in the Energy Vision felt this to be a "warm shower" of interest and consent.

Taken seriously was also the ability of the initiative to successfully influence (government) decisions at various levels. The initiators established good relationships with officers in the provincial government and a relevant federal ministry and these expressed their support for the initiative on various occasions.

The example of Murau provides many illustrations of local environmental activities that required support from and changes at provincial or national levels (e.g. changes in provincial or national regulation and support structures). On the other hand, the local activities were used as exemplars and a source of legitimacy for activities which aim at a transition of the energy system at a national, European or global level. Partly due to the importance of such local demonstrations for the general cause of energy transitions, pioneering initiatives especially in Austria often get directly supported by allies in federal ministries via specific funding schemes or by awarding publicity winning prizes etc. In our case, the mutual support of "local" discourse and action on one side and of "provincial and national" discourse and action on the other side, actually seems to be of crucial importance. To what extent such initiatives in fact do currently destabilize the energy regime will only be assessable in retrospect. But there are remarkable shifts of discourses to be observed internationally. This is particularly true in the EU and in Asian countries, where the energy policy discourse was dominated over many decades by experts who did not at all allow deviation from the ideal of a centralized, fossil-fuel-based energy system. This dominance came to an end at around the turn of the century, making way for a much more pluralistic debate including visions of sustainability and regional energy autarky.

4. Conclusions: Possible Roles of Cities and Regions in Regime Change

As we discussed in Section 2, it is difficult to attribute the notions of regimes and niches to empirical cases of local variation in cities and regions. Against our empirical observations, the notion of regimes that collectively, and globally, resist change appears to suggest too much homogeneity and consistency. While there certainly are dominant technologies, practices and institutions at a national or even global level framing the way energy is generated and used, there is considerable variation also of "regime structures" at the regional/local level. This can be substantiated by looking at the spatial variation of regimes in "high resolution".

These variations are smaller in closely coupled infrastructure networks, such as the electricity system, but larger for other components of the energy regime such as heat provision. We see, for example, similarities with the remnants of older stable regimes such as the tradition of wood-heating in Austrian rural areas like Murau. The fact that many farmers still rely on wood from their own forests (a particularity of the Austrian agricultural structure) or on personal and "untraded" relations with forest owners is reminiscent of a "pre-modern" energy regime, even though this "culture of wood heating" to an extent served as a basis for modern renewable wood-heating systems (Rohracher, 2002). Regimes are thus full of inconsistencies, irregularities and conflict which may provide "anchors" and allies for emerging socio-technical configurations. Moreover, radical shifts are usually impossible to identify from a close view, where changes are rather gradual and patchy – in contrast to the long-term historic perspective transition case studies often take (see, e.g. Geels, 2005; Geels & Kemp, 2007).

Similarly, problems arise in applying the "niche" notion to our case. Although there are interesting examples of protected spaces where novel technologies are conceived, tested and contextualized, many of the environmental projects in Murau are not so much about experiments with new technologies in protected spaces. Most of the activities focused on providing supportive frameworks and socio-technical configurations for the broader application of given technologies, such as local heating networks based on woodchips and biogas, small hydropower, a wind turbine as well as many installations for solar water heating.

Rather than just providing favourable contexts for and accumulating niche experiments, cities or regions appear to bridge the niche-regime divide and provide "social contexts" (actor networks, institutions, complementary technological structures) "to integrate" and implement socio-technical configurations which differ from the dominant regime and may be important for long-term transition processes. The aim is not so much to strategically create technological niches and promote alternative configurations on the basis of an evolutionary understanding of technological change but rather to demonstrate the feasibility of new configurations that differ from those usually entrenched in the regime—technical alternatives combined with matching institutions, practices, etc. The main task of the initiative is to adapt these alternative configurations and implement them in the given socio-economic conditions (framework conditions, priorities, support structures, etc.) of a particular territory as a demonstration of feasibility.

Local actor networks can mobilize resources specifically accessible to them like organizational capabilities, funding, social capital and institutional adaptation. As a consequence, the systems of provision that are realized in cities and regions can deviate quite substantially from dominant regime characteristics. And when favourable circumstances such as a crisis of the entrenched regime or other "windows of opportunity" emerge, such deviations can play an important role in undermining the authority and belief in the dominance of the entrenched regime.

In order to understand the motivation of actors to play a pro-active role in such a process, we need to look at the dynamics between local node and global networks in a more detailed way from the perspective of multi-level governance. A remote district can gain a lot of recognition if it develops into a showcase for a "sustainable energy future", and there are quite important social and economic benefits to be gained from such a positioning, e.g. in a deprived region that is putting much hope on tourism, that is prone to outmigration, etc. On the level of single individuals, many entrepreneurial opportunities seem to arise with the need for networking, intermediation, transfer of knowledge, promotion of policy ideas, etc., which can be beneficial, e.g. in terms of social status.

The capabilities and limitations for such change processes originating at a local level in turn depend on the embedding of cities and regions in multi-level governance systems (which has to be kept apart from the "multi-level perspective") and also on the room for manoeuvring within incumbent socio-technical regime structures. Many regime structures are of course beyond reach for local actors. However, being part of a multi-level governance system not only restricts the capacity of e.g. local governments to achieve change at a more radical level, but also opens up additional opportunities to solicit and embed change processes at other scale levels.

We tried to demonstrate with our example that it is worth paying more attention to the interplay of local and non-local discourses. Only an analysis that focuses on the dynamic and mutually beneficial interplay of (a) local demonstrations and (b) (rather) global networks supporting particular visions of a socio-technical transition, can enable a comprehensive assessment of potential transition dynamics, particularly because of their potential role in the legitimization and entrenchment of alternative socio-technical configurations. While the growth and stabilization of niches have been primarily conceptualized as niche-accumulation dynamics in transition studies, the process of legitimization and its spatial dynamics have so far received little attention. A demonstration of the socio-economic feasibility and desirability of particular (originally local) regime deviations can, however, provide crucial opportunities for mobilizing authority and other resources for these alternatives, which might before have been marginal compared to the entrenched regime. This is the rationale behind and motivation for the setting up and support of hundreds of (bio)energy regions, eco-villages and model cities worldwide. We have illustrated with our example some of the mechanisms and processes through which regional governance can potentially have a broader impact on energy system transitions. We showed how local demonstrations and actors in favour of socio-technical transitions on the national and international level have a mutual interest in supporting each other. And that social, institutional and discursive particularities of different places can foster the emergence of heterogeneous social networks that are capable of mobilizing funds, attention and credibility for ambitious visions of change.

We have come to the conclusion that despite many contingencies and restraints, cities and regions can provide a specific and potentially powerful social context for such deviations from the dominant regime. Not least due to the inconsistencies, frictions and often asynchronous developments within existing regime structures opportunities can emerge to stabilize and embed such deviations. Places can emerge as central arenas not only for vision building and particular learning processes but also for demonstrating the feasibility of alternative visions and hence for the accumulation of credibility, which can strengthen actor networks and institutions globally. As far as these developments result in the emergence or maturation of new social norms, they also impact on regimes via the landscape level. Change in this context, however, can be initiated by factors other than just niche-regime interactions. It can also be driven by the specific dynamics and restraints of multi-level governance interactions and by contingent regional contexts as well as by particular expectations and actor networks.

What we thus see in such cases is the inter-linkage of multi-scalar governance systems and multi-level socio-technical structures. Both have their particular dynamics and logics of change, but these dynamics are interlinked and mutually shape each other. Just as we need to consider socio-spatial dynamics across different scales to explain the partial stabilization of local alternatives to the incumbent energy regime, we need an understanding of the dynamics of socio-material infrastructures to better explain dynamics of urban or regional change processes. Our empirical case emphasizes the importance of these interlinkages of multiple levels of socio-technical structuration (such as the dynamic interplay of niches and regimes) and multiple socio-spatial scales of governance (such as the interplay of regional, national and international governance systems) to more fully comprehend the dynamics of energy system transformation. At a conceptual level, however, these couplings still wait for a convincing integration – an enterprise which has to be based on the comparison and evaluation of a broader range of empirical cases in different contexts.

In terms of policy implications, these observations seem to support many ongoing initiatives to intentionally create pressure on regimes that are recognized as unsustainable. The focus of attention might have to be shifted, though, away from the creation of ever more experiments and stimulations for the market introduction of technologies towards strategic interventions in support of emerging discourses and their institutionalization. Moreover, the urban or regional level deserves to receive more policy support for creating coherent and integrated alternative solutions which go beyond isolated demonstration projects. Such solutions apparently do have the potential to have an impact on discourses, strategies and the formation of actor alliances at national or international scale or to horizontally link-up in green city networks or similar initiatives. However, based on our investigations we can merely identify such mechanisms for the shaping of transition dynamics—what their ultimate impact on the transformation of energy systems will be still remains to be seen.

In terms of an agenda for research on these sustainability transitions, we insist on our call for a space-conscious analysis of ongoing and past challenges to incumbent regimes and the maturation of an analytical framework that was better able to guide the study of such cases (also in comparison). We tried to include in our analysis, both locally specific conditions for regime-deviations and the agency of local actors in building social networks. We also looked at the intermediation between different sectors of society and different governance levels, which requires a high (spatial) resolution, but seems to be essential for an understanding of the discursive dynamics around alternative constellations, which we believe to be important for the (de-)stabilization of socio-technical regimes.

Notes

1. For the distinction of place and territory we follow Jessop *et al.* (2008).
2. The empirical data has been gathered mainly in the context of two research projects conducted in 2005–2009. These projects have been funded by the Austrian Federal Ministry of Transport, Innovation and Technology, which is gratefully acknowledged. We mainly build our case on semi-structured interviews with local professionals and voluntary activists and a series of local workshops. Moreover, one of the projects comprised a joint learning process among three prominent Austrian "Energy Regions" (Späth *et al.*, 2007, 2010).
3. We are very sceptical that a simple transfer of assumptions regarding the dynamics of socio-technical regimes from this global or national level to a local or regional level will bear fruit, as is assumed in some parts of the Dutch transition management literature (Loorbach, 2007).
4. For a more detailed discussion of this case, see Späth and Rohracher (2010).
5. Besides these important three objectives, there are two more which, however, largely overlap with the three documented here.

References

Alber, G. (2009) Energieregionen in Österreich, in: D. Keppler, H. Walk, E. Töpfer & H.-L. Dienel (Eds) *Erneuerbare Energien ausbauen!*, pp. 131–148 (München: Oekom).

Bache, I. & Flinders, M. (2004) Themes and issues in multi-level governance, in: I. Bache & M. Flinders (Eds) *Multi-level Governance*, pp. 1–14 (Oxford: Oxford University Press).

Bauriedl, S. (2009) Adaptive capacities of European city regions in climate change: On the importance of governance innovations for regional climate policies, Climate 2009, Online Conference. www.climate2009.net (retrieved 5 November 2009).

Benz, A. (2004) Leistungswettbewerbe in der regionalen Raumentwicklungspolitik, DISP (157), pp. 4–10.

Benz, A. & Eberlein, B. (1999) The Europeanization of regional policies: Patterns of multi-level governance, *Journal of European Public Policy*, 6(2), pp. 329–348.

Böcher, M., Krott, M. & Tränkner, S. (2008) *Regional Governance und integrierte ländliche Entwicklung—Ergebnisse der Begleitforschung zum Modell- und Demonstrationsvorhaben "Regionen Aktiv"* (Wiesbaden: VS-Verlag).

Braun, M. & Santarius, T. (2008) *Climate Politics in the Multi-level Governance System. Emissions Trading and Institutional Changes in Environmental Policy Making*. Wuppertal Papers. Wuppertal, Wuppertal Institute for Climate, Environment and Energy, 172, p. 37.

Bulkeley, H. (2005) Reconfiguring environmental governance: Towards a politics of scales and networks, *Political Geography*, 24(8), pp. 875–902.

Bulkeley, H. & Betsill, M. (2005) Rethinking sustainable cities: Multilevel governance and the 'urban' politics of climate change, *Environmental Politics*, 14(1), pp. 42–63.

Bulkeley, H. & Kern, K. (2006) Local government and the governing of climate change in Germany and the UK, *Urban Studies*, 43(12), pp. 2237–2259.

Campbell, J. L. (1998) Institutional analysis and the role of ideas in political economy, *Theory and Society*, 27(3), pp. 377–409.

Coenen, L., Raven, R. & Verbong, G. (2010) Local niche experimentation in energy transitions: A theoretical and empirical exploration of proximity advantages and disadvantages, *Technology in Society*, 32(4), pp. 295–302.

Coutard, O. & Guy, S. (2007) STS and the city: Politics and practices of hope, *Science Technology Human Values*, 32(6), pp. 713–734.

David, P. A. (1985) Clio and the economics of QWERTY, *American Economic Review, Proceedings*, 75(2), pp. 332–337.

Elbe, S., Kroës, G., Benz, A., Lukesch, R., Weiβ, K., Böcher, M., Krott, M., Meincke, A., Middelmann, U., Payer, H., Rabenau, J. & Tränkner, S. (2007) *Begleitforschung Regionen Aktiv—Syntheseberichte* (Göttingen: Universitätsverlag).

Geels, F. W. (2004) From sectoral systems of innovation to socio-technical systems: Insights about dynamics and change from sociology and institutional theory, *Research Policy*, 33(6–7), pp. 897–920.

Geels, F. (2005) *Technological Transitions and System Innovations—A Co-evolutionary and Socio-Technical Analysis* (Cheltenham: Edward Elgar).

Geels, F. W. & Kemp, R. (2007) Dynamics in socio-technical systems: Typology of change processes and contrasting case studies, *Technology in Society*, 29(4), pp. 441–455.

Geels, F. & Raven, R. (2006) Non-linearity and expectations in niche-development trajectories: Ups and downs in Dutch biogas development (1973–2003), *Technology, Analysis Strategic Management*, 18(3/4), pp. 375–392.

Geels, F. W. & Schot, J. (2007) Typology of sociotechnical transition pathways, *Research Policy*, 36(3), pp. 399–417.

Graham, S. & Marvin, S. (2001) *Splintering Urbanism. Networked Infrastructures, Technological Mobilities and the Urban Condition* (London: Routledge).

Grin, J., Rotmans, J. & Schot, J. (2010) *Transitions to Sustainable Development—New Directions in the Study of Long Term Transformative Change* (New York: Routledge).

Guy, S. & Marvin, S. (1996) Disconnected policy: The shaping of local energy management, *Environment and Planning C: Government and Policy*, 14(1), pp. 145–158.

Hajer, M. A. (1995) *Discourse Analysis. The Politics of Environmental Discourse. Ecological Modernization and the Policy Process*, pp. 42–72 (Oxford: Clarendon Press).

Hajer, M. & Wagenaar, H. (2003) Introduction, in: M. Hajer & H. Wagenaar (Eds) *Deliberative Policy Analysis*, pp. 1–30 (Cambridge: Cambridge University Press).

Hodson, M. & Marvin, S. (2009) Cities mediating technological transitions: Understanding visions, intermediation and consequences, *Technology Analysis & Strategic Management*, 21(4), pp. 515–534.

Hoogma, R., Weber, M. & Elzen, B. (2005) Integrated long-term strategies to induce regime shifts towards sustainibility: The approach of strategic niche management, in: M. Weber & J. Hemmelskamp (Eds) *Towards Environmental Innovation Systems*, pp. 209–236 (Berlin: Springer).

Jessop, B., Brenner, N. & Jones, M. (2008). Theorizing sociospatial relations. *Environment and Planning D: Society and Space*, 26(3): 389–401.

Knill, C. & Liefferink, D. (2007) *Environmental Politics in the European Union* (Manchester: Manchester University Press).

van Lente, H. (2000) Forceful futures: From promise to requirement, in: N. Brown, B. Rappert & A. Webster (Eds) *Contested Futures—A Sociology of Prospective Techno-Science*, pp. 43–64 (Burlington: Ashgate Publishing Company).

Loorbach, D. (2007) *Transition Management—New Mode of Governance for Sustainable Development* (Utrecht: IPG Books).

Loorbach, D. (2010) Transitions management for sustainable development—a pescriptive, complexity-based governance framework, *Governance*, 23(1), pp. 161–183.

Marks, G. & Hooghe, L. (2004) Contrasting visions of multi-level governance, in: I. Bache & M. Flinders (Eds) *Multi-level Governance*, pp. 15–30 (Oxford: Oxford University Press).

Mirumachi, N. & Van Wyk, E. (2010) Cooperation at different scales: Challenges for local and international water resource governance in South Africa, *Geographical Journal*, 176(1), pp. 25–38.

Monstadt, J. (2007) Urban governance and the transition of energy systems: Institutional change and shifting energy and climate policies in Berlin, *International Journal of Urban and Regional Research*, 31(2), pp. 326–343.

Morgan, K. (2004) Sustainable regions: Governance, innovation and scale, *European Planning Studies*, 12(6), pp. 871–889.

Moser, P., Kucharczak, L. & Hoppenbrock, C. (2009) How to achieve renewable energy regions and advance sustainable development: Integrated models and processes in Germany, in: P. Droege (Ed.) *100% Renewable—Energy Autonomy in Action*, pp. 173–186 (London: Earthscan).

Radzi, A. (2009) 100% Renewable champions—international case studies, in: P. Droege (Ed.) *100% Renewable—Energy Autonomy in Action*, pp. 93–166 (London: Earthscan).

Rip, A. & Kemp, R. (1998) Technological change, in: S. Rayner & E. L. Malone (Eds) *Human Choice and Climate Change: Resources and Technology* (vol. 2, pp. 327–399 (Columbus, OH: Batelle Press).

Rohracher, H. (2002) A socio-technical mapping of domestic biomass heating systems in Austria, *Bulletin of Science, Technology & Society*, 22(6), pp. 474–483.

Rohracher, H. & Späth, P. (2009) The fragile dynamics of urban energy system transitions: The eco-cities of Graz and Freiburg in retrospect. International Roundtable Conference on 'Cities and energy transitions: past, present, future', Autun, France.

Schot, J. & Geels, F. W. (2008) Strategic niche management and sustainable innovation journeys: Theory, findings, research agenda, and policy, *Technology Analysis & Strategic Management*, 20(5), pp. 537–554.

Smith, A. (2007a) Emerging in between: The multi-level governance of renewable energy in the English regions, *Energy Policy*, 35(12), pp. 6266–6280.

Smith, A. (2007b) Translating sustainabilities between green niches and socio-technical regimes, *Technology Analysis & Strategic Management*, 19(4), pp. 427–450.

Smith, A., Stirling, A. & Berkhout, F. (2005) The governance of sustainable socio-technical transitions, *Research Policy*, 34(10), pp. 1491–1510.

Späth, P. (2008) 'Energy Regions': Guiding visions that work?—The influential alignment of actors around visions of long-term socio-technical change, in: *2008 Berlin Conference: Long-Term Policies: Governing Social-Ecological Change*, Berlin, FFU.

Späth, P. & Rohracher, H. (2010) 'Energy regions': The transformative power of regional discourses on socio-technical futures, *Research Policy*, 39(4), pp. 449–458.

Späth, P., Koblmüller, M., Kubeczko, K., Faber, F., Bärnthaler, J., Bergmann, H., Luttenberger, C. & Breisler, A. (2007) 'EnergieRegionen': Wirksame Leitbildprozesse und Netzwerke zur regionalen Gestaltung sozio-technischen Wandels—Endbericht Wien, BMVIT: Berichte aus Energie- und Umweltforschung, Band 29/07.

Späth, P., Schreuer, A., Koblmüller, M., Bärnthaler, J., Hannesschläger, M. & Luttenberger, C. (2010) *Entscheidungshilfen für Förderer und KoordinatorInnen von EnergieRegionen. Endbericht* (Graz, Wien: BMVIT).

STRN Steering Group (2010) A mission statement and research agenda for the Sustainability Transitions Research Network. Available at http://www.transitionsnetwork.org/about (retrieved 29 November 2012).

Truffer, B. (2008) Society, technology and region: Contributions from the social study of technology to economic geography, *Environment and Planning A*, 40(4), pp. 966–985.

Wenger, E. (2000) Communities of practice and social learning systems, *Organization*, 7(2), pp. 225–246.

Wissen, M. (2009) Contested terrains: Politics of scale, the national state and struggles for the control over nature, *Review of International Political Economy*, 16(5), pp. 883–906.

Index

Note: Page numbers in **bold** type refer to figures
Page numbers in *italic* type refer to tables
Page numbers followed by 'n' refer to notes